Petrini & Petrini

Identifying moulds

Liliane E. Petrini
Orlando Petrini

Identifying Moulds

A Practical Guide

with 37 figures und 12 tables

J. Cramer
in der Gebr. Borntraeger Verlagsbuchhandlung 2013

Petrini, L.E. & Petrini, O.: Identifying Moulds

Authors' address:

Liliane E. Petrini
Orlando Petrini
Via al Perato

6932 Breganzona
Schweiz

liliane@petrininet.ch

Pictures on the back cover (from top to bottom):
Melanospora fusispora: Fruiting body; *Ulocladium chartarum*: conidiophore, conidia; *Trichothecium roseum*: conidiophore, conidia; *Mucor plumbeus*: sporangia, columella, sporangiophores.

Please send comments to mail@schweizerbart.de

ISBN 978-3-443-50038-2
Information on this title: **www.borntraeger-cramer.de/9783443500382**

This title is also available in German, information on the German title:
www.borntraeger-cramer.de/9783443500351

Publisher: Gebr. Borntraeger Verlagsbuchhandlung
Johannesstr. 3A
70176 Stuttgart, Germany
mail@borntraeger-cramer.de
www.borntraeger-cramer.de

∞ Printed on permanent paper conforming to ISO 9706-1994
Printed in Germany by strauss offsetdruck gmbh, 69509 Mörlenbach

Contents

I. Introduction

1. Of fungi and man

For most people, fungi are nothing else but those mushrooms one can pick up in the forest: some are edible, some poisonous, some are more common than others, but all share a similar shape, having a stem and a cap with gills or tubes. Invariably, people are most likely to mention the fly agaric (*Amanita muscaria*) or the button mushroom (*Agaricus campestris*) as "the" fungal representatives. Yet the basidiomycetes, to which both fly agaric and button mushroom belong, are just a small selection of all known fungi and comprise a very large number of species that are extremely diverse – for instance, in addition to the "classical" mushrooms (Agaricomycotina), the Basidiomycota include the plant pathogenic rusts (Puccinomycotina) and smuts (Urediniomycotina), the jelly fungi, the stinkhorns, as well as several less known groups.

The very large and heterogeneous kingdom Fungi is composed of eukaryotic organisms and is clearly separate from bacteria, plants and animals. Fungal cells have mainly cell walls that contain chitin and fungi are physiologically quite different from plants and animals. The Kingdom Fungi includes several taxonomic units, listed in Chapter III. Some fungi are one-celled, others many-celled: in fact, the largest organism on earth appears to be the basidiomycete fungus *Armillaria ostoyae*, a very common, feared tree pathogen: the mycelium produced by one individual has been shown to occupy 965 hectares of soil in Oregon's Blue Mountains!

Most fungi form a vegetative mat called mycelium, composed of thread-like, mostly branching filaments termed hyphae. In the mushrooms, the cap-like structures one sees in the woods are the fruiting bodies, the so-called "carpophores" – comparable to the fruits of a tree: in them, reproductive structures (the basidia) form propagules (the spores) that can be dispersed by wind, water or insects: similarly to plant seeds, they can germinate and produce mycelia that colonise new substrates. Hyphae are usually very small and only the mycelium can sometimes be observed as a whitish or coloured mat covering the surfaces colonised by the fungus. Mostly, however, the vegetative fungal structures are inconspicuous and even the carpophores are microscopic.

Fungi are everywhere: they may grow in soil, on dead organic material or live symbiotically in or on plants, animals or other fungi. They are often noticed only when fruiting or causing disease symptoms on their hosts.

Fungi are known to influence their environment and to induce significant physiological changes in it. In turn, the ecosystem greatly influences fungal growth, exerting a strong selective pressure. Ecologically, fungi play a crucial role in nutrient cycling and exchange. Their importance as a direct source of food is undisputable: mushrooms and truffles are very appreciated; yeasts are used as leavening agents for bread and in the wine and beer fermentation, and several moulds are indispensable for the fermentation of various other food products. In the medical field, antibiotics and many other drugs are extracted from fungi (e.g. penicillin from *Penicillium*; the immunosuppressant Cyclosporin A from *Tolypocladium*; psychotropic compounds and alkaloids from Clavicipitaceae) and many moulds are used in industrial applications, e.g. for the production of enzymes and detergents. Finally, some fungi have found a place as biological pesticides to control weeds and harmful insects.

While several fungi are beneficial, many can be harmful to humans, animals and plants. Some moulds can grow and damage manufactured materials and buildings: high fungal loads can be present in damp buildings that have not been properly restored, particularly after serious water damage. Approximately 300 fungal species are known to cause infections ("mycoses") in humans and animals. Inhalation of mycotoxins or spores can cause severe allergic reactions and asthma in debilitated and predisposed subjects. Allergic reactions caused by fungi have been reported from malt and cheese factories, bakeries, mills, and composting facilities, all sites where high loads of fungal spores may be present.

Economically most relevant, however, are the crop losses caused by some plant pathogenic fungi (e.g. the rice blast disease by the ascomycete *Magnaporthe grisea* or black rot of grape by *Guignardia bidwellii*). Finally, several moulds are involved in food spoilage and mycotoxin production – they, too, are economically important. So far some 400 different mycotoxins, produced by approximately 350 fungal species, have been described. The most prominent toxin producing fungi belong to the genera *Aspergillus*, *Fusarium*, and *Penicillium*.

Approximately 70,000 fungal species are known, but the total number of specific fungal taxa is estimated to be as high as 1.5 millions. Among them, moulds are by far the economically most important.

Moulds grow usually saprobically on organic substrates and may cause modifications of the substrate's appearance, texture, and sometimes (in the case of food) taste. "Mould" is no taxonomic, physiological or biochemical term: moulds are fungi, but not all fungi should be called moulds. Systematically, they belong to several taxonomic groups. The term is not linked to the practical significance of the organism either. Moulds can cause disease or food spoilage, some play an important role in biodegradation, others in the production of food (e.g. *Penicillium roqueforti* in blue cheese), antibiotics (*Penicillium* spp. as penicillin producers), enzymes (cellulases produced by *Trichoderma* spp.), or organic substances (citric acid by *Aspergillus niger*), whereas some may be opportunistic, human and animal pathogens.

2. How to use this book

This laboratory guide is intended to be an introduction to the identification of the most common and important genera of moulds. It considers only fungi that fruit readily in pure culture, are commonly found in the environment, are common plant pathogens or are involved in food spoilage, or may be isolated from human and animal tissues as opportunistic pathogens. It has been written for students and lab technicians and should help them find their way in the complicated taxonomic world of microscopic fungi. Genera not included here are probably rather uncommon; in this case, more specialised monographs will be indispensable for their identification – they are referred to in "Common mould genera" on page 89.

No previous knowledge of fungi is expected in order to use this book. Sections "Fundamentals of mycology" on page 5 and "Taxonomy" on page 15 give some basic information on mycology as well as brief descriptions of the most important taxonomic schemes. The section "The identification of moulds" on page 25 describes briefly the methods used for the identification of moulds and other fungi, whereas Sections V–VII provide practical advises on handling, safety issues and steps to be taken during the identification work. They also contain practical information on culture media and morphological characters typical of the different fungal groups.

Sections VIII and IX are the core of this manual. Section VIII provides a key to the most common mould genera, while Section IX is a catalogue of the genera keyed out in Section VIII, with reference to more specific identification literature.

The mycological taxonomic literature is unfortunately very rich in technical definitions. We provide a glossary in Section X that includes some of the most relevant mycological terms.

Sections XI–XII give a very brief introduction to the ecology of fungi in general and to the significance of fungi for public health.

Those who are interested only in the identification of moulds and already have some working knowledge of mycology will probably use only Sections VIII and IX. The other chapters are intended to provide a very basic introduction to mycology and fungal identification – they are, by no means, in-depth treatises, but contain the necessary bibliographic references to gain additional knowledge on the subjects treated.

II. Fundamentals of mycology

1. What are fungi?

Fungi, formerly considered members of the plant kingdom, have now been partly included in a separate kingdom (Fungi: see section III on page 15). A small number has been transferred to the Protista. Fungi and fungus-like Protista show morphological, ecological and biochemical features that distinguish them from other organisms.

These are briefly summarised in Tab. II.1.1.

Table II.1.1: Features of fungi and fungus-like Protista

Nutrition	Heterotrophic organisms (no photosynthesis). Direct absorption of nutrients.
Thallus	The vegetative part of a fungus; in the Protista mainly a protoplast or a fully developed structure with distinct cell walls, composed at the beginning by several undifferentiated vegetative units that reproduce and absorb nutrients independently. The thallus may later differentiate into distinct organs. In most cases the thallus is not mobile, but some fungus-like Protista can produce zoospores or myxoamoebae.
Cell wall	Usually present, normally containing chitin (cellulose may be present in some groups – regularly in Oomycota and in some Ascomycota).
Nucleus	Eukaryotic (enclosed by a membrane), homo- or heterokaryotic, dikaryotic, haploid or diploid (diploidy is usually limited in time).
Reproduction	Sexual and asexual.
Spores	Non-motile and motile spores (zoospores)
Fruiting bodies	(Also called sporocarps) very small to visible with the naked eye, with limited tissue differentiation.
Occurrence	As symbionts, saprobes, parasites or hyperparasites.
Distribution	Worldwide.

2. Morphology and reproduction

Fungi and fungal Protista share properties that may be found also within other eukaryotes. Not necessarily common morphological or biochemical characters are indicative of a close phylogenetic relationship: the same characters may be present in taxonomically unrelated taxa. The most important morphological characters are presented in Figs. II.2.1–3.

2.1 Vegetative organs

2.1.1 The thallus (Fig. II.2.1)

Spores germinate and produce a germ tube that will eventually elongate and branch to form a hypha. Yeasts usually reproduce by budding and some fungus-like Protista may produce zoospores in closed organs called sporangia.

The thallus is the undifferentiated vegetative tissue of a fungal organism and may have different structural characteristics depending on the taxonomic position of the fungus.

- The mycelium is the vegetative part of a fungus and is composed of a mass of thread-like, often branching hyphae. These are long, branching filamentous structures that may be septate (with cross-walls: Ascomycota and Basidiomycota) or aseptate (without cross-walls: Zygomycota). Hpyhae absorb nutrients from the substrate. Within the hyphae the nutrients are transformed and energy is produced. In septate hyphae, nutrients pass the septa through central pores.Within the thallus, hyphae can differentiate and take up different tasks. Hyphae immersed in the substrate absorb nutrients and metabolise them; aerial hyphae are mainly specialised in rapid growth and substrate colonisation. Reproduction often takes place on aerial hyphae; stolons (runners) are aerial hyphae that are used by the fungus to rapidly colonise new substrates. They often form root-like structures (rhizoids) at the point of contact with the substrate. Rhizoids are special organs that are able to absorb nutrients.

- Yeast colonies are composed of single, totipotent cells, each of them able to absorb nutrients and reproduce by budding or cell fission. Some yeasts (e.g. *Candida*, *Torulopsis*) may produce pseudohyphae and in some cases true hyphae. Pseudohyphae are basically chains of yeast cells that differ from true hyphae by being constricted at the septa. They can be recognised also by their terminal cells that are usually smaller than the other cells.

- Some filamentous fungi (e.g. *Rhizopus oryzae*, Zygomycota) may form budding cells and grow yeast-like under specific ecological conditions. Such organisms are called dimorphic.

- Dimorphic fungi may grow either yeast-like or form hyphae. Switching from the budding yeast to the filamentous form is mainly triggered by ecological factors. Several dimorphic fungi are facultative human or animal pathogens: the parasitic form is mainly yeast-like, while the saprobic phase is filamentous.

Dimorphism may be monofactorial or multifactorial. For instance, the two human pathogens *Blastomyces dermatitidis* and *Paracoccidioides brasiliensis* grow as filamentous organisms at temperatures below 37 °C and yeast-like above, whereas *Candida albicans* grows yeast-like at high sugar concentrations or in the presence of proteins with high sulphide content. In rare cases the switch from yeast-like to filamentous growth is triggered by a combination of factors, as is the case for the human pathogen *Histoplasma capsulatum*, which switches from the mycelial to its yeast form when the temperature reaches 37 °C in the presence of high sugar concentrations and proteins with high sulphide content. Some Mucorales, *Penicillium* and *Aspergillus* species, grow yeast-like under anaerobic or microaerobic conditions.

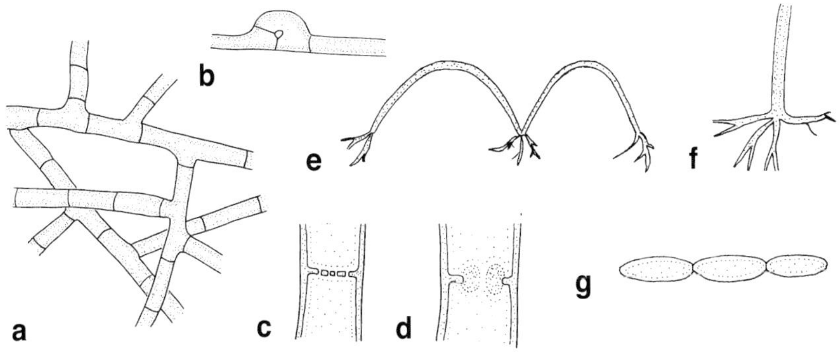

Fig. II.2.1. Vegetative organs

a. Mycelium composed of septate, branched hyphae. **b.** Hyphal septum with clamp connection (only in Basidiomycota). **c, d.** Pores: **c.** Micropores. **d.** Dolipores (only Basidiomycota). **e.** Stolons (runners). **f.** Rhizoids. **g.** Pseudohypha.

2.1.2 Vegetative structures (Fig. II.2.2)

Special organs (Tab. II.2.2.1) are often produced in the thallus following ecological or physiological changes.

- **Aleurospores** are thick-walled, vegetative, resting spores formed at the hyphal tips. They are resistant to adverse ecological and physiological factors (desiccation, high temperatures, etc.), thereby ensuring the survival of the fungus. Morphologically they are almost indistinguishable from chlamydospores (Figs. VIII.5d, VIII.13f).

Tab. II.2.2.1: Vegetative organs of the thallus

Physiological or ecological function	**Fungal structure**
Germination	Germ tubes, budding cells.
Infection	Infection hyphae.
Spread and increase of the fungal mass	Perforating hyphae, stolons, microhyphae, aerial hyphae, rhizomorphs.
Survival	Chlamydospores, aleurospores, sclerotia.

- **Chlamydospores** are thick-walled, often melanised hyphal cells of very variable shape (Fig. VIII.14b). They are resting, long-living and resistant organs able to store nutrients and protect the fungus' genetic material from ecological and chemical damage. They are hardly different from aleurospores, from which they can be distinguished mainly by the site of formation (not on hyphal tips) and their very variable and irregular shapes.
  In the Zygomycota, chlamydospores are mainly formed endogenously within the hyphae or the sporangia; in the Ascomycota and Basidiomycota they may be formed either endogenously or exogenously. Sometimes, Zygomycota may produce rather thin-walled chlamydospores (called **gemmae**) that separate from the mycelium and may in turn reproduce by yeast-like budding.

- **Hairs**, **bristles** and **setae** (bristle-like hairs) are sometimes formed by some mould species. The morphological differences among the three are subtle: hairs are usually slightly thinner than hyphae, whereas bristles and setae are thicker and often tapering; bristles and setae are morphologically almost indistinguishable. Hairs, bristles or setae differ from normal hyphae also by their colour, being usually darker. Their physiological function is not clear – some authors have hypothesised that their purpose is to protect the fungus from grazing by insects. Hairs and setae are often found in or on fruiting bodies of ascomycetes and their anamorphs (e.g. on *Chaetomium* ascomata). They may be characteristic of selected fungal taxa and often help in their identification, as is the case for the plant pathogenic anamorphic genus *Colletotrichum* (Fig. VIII.8d).

- **Infection** or **perforating hyphae** are penetration organs. They are able to perforate the host cell walls mechanically or enzymatically. Plant pathogens, for instance, can digest the host's cuticula and cell walls with enzymes such as pectinases, cutinases, cellulases or other hydrolases. Defensive responses by hosts include callus formation and necrosis in plant infection, whereas in animals and humans mould penetration will elicit inflammation and immunological responses.

- **Microhyphae** are extremely thin, penetrating hyphae (less than 0.5 μm in diameter) formed by some plant pathogenic fungi.

- **Rhizomorphs** are strands of organised hyphae that closely resemble thin roots. They can be up to 1 cm thick and their structure may be already quite differentiated, with an outer cortex-like and an inner mycelial layer. Rhizomorphs are mainly absorption and transport organs, at the same time allowing a rapid substrate colonisation by the fungus. The rhizomorphs of some basidiomycetes (eg. *Armillaria mellea* or *Serpula lacrymans*) are able to penetrate wood and building walls and may reach several meters in length.

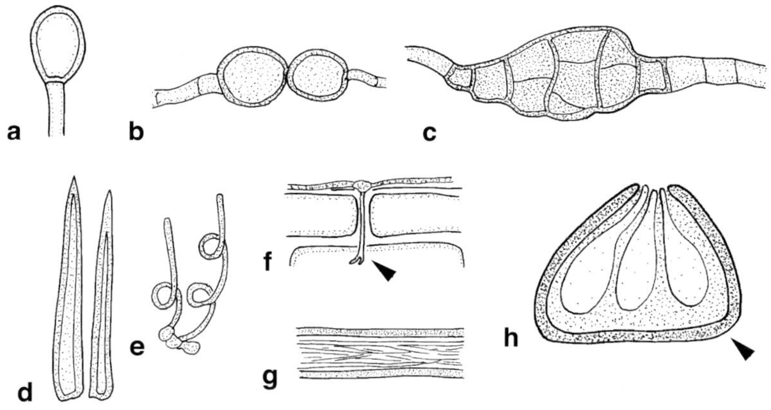

Fig. II.2.2. Organs of the thallus

a. Aleurospore. **b.** One-celled chlamydospores. **c.** Multicellular chlamydospore. **d.** Setae (bristles). **e.** Hairs. **f.** Perforating (infection) hypha (arrow). **g.** Rhizomorph. **h.** Stroma (arrow) with fruiting bodies inside.

- **Sclerotia** are hard aggregations of hyphae, mostly of cellular structure (the "medulla") and with a superficial, melanised rind (the "cortex"), functioning as a resting body (Fig. VIII.12b, c).

- **Stromata** (sing. stroma) are compact structures composed of hyphae, often with a cellular, parenchymatic appearance, in or on which one or more fruiting bodies are seated. Stromata may occur in asexual and sexual fructifications.

2.2 Reproductive organs

Careful study of the morphology of a mould is a crucial first step towards its identification: this includes examining and describing its vegetative and reproductive organs.

Fungi reproduce by means of propagules. These are called spores. Spores may be produced either asexually (no meiosis – only by mitosis) or sexually (meiosis following fusion of nuclei).

Fungi are not only genetically but also morphologically extremely diverse. They may be pleomorphic, meaning that they may assume different forms and shapes, and reproductively form only the sexual or the asexual form, or both. The whole fungus (i.e., all possible morphological and genetic expressions of a taxon) is called the holomorph. This includes its asexual form (the so-called anamorph, from "anatomic morphology") and its sexual form (the teleomorph – "perfect" form). A fungal species reproducing only asexually is sometimes called "mitosporic", whereas those taxa forming a teleomorph are sometimes referred to as meiosporic fungi. To complicate things further, a fungus can sometimes form several anamorphs, all known by different names.

Accordingly, the sexual and the asexual forms of the same fungal species are often known by two different names. One designates the asexual, the other the sexual morph – both together form the holomorph. For instance, the names *Aspergillus hollandicus* and *Eurotium amstelodami* define the same ascomycetous fungus: *A. hollandicus* is the name of the (asexual) anamorph, whereas *E. amstelodami* describes the (sexual) teleomorph. *Aspergillus* is therefore the anamorph of *Eurotium* and, correspondingly, *Eurotium* the teleomorph of *Aspergillus* (Fig. II.2.3). New nomenclatural rules call for the use of only one name, but this will most likely take some time before being fully established.

In the absence of one of the two forms, a taxon is known either by its anamorph or its teleomorph name. Mycologists are currently trying to simplify this confusing situation by using molecular biology techniques – but it will take some time to sort out all known and unknown cases and to call a fungus by a single name.

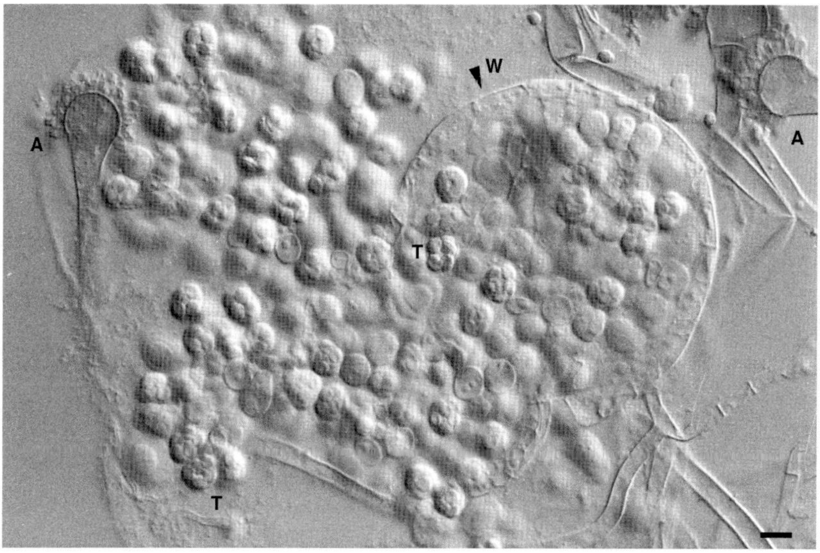

Fig. II.2.3 ***Anamorph-Teleomorph connection:*** **Aspergillus hollandicus** ***and*** **Eurotium amstelodami**

Holomorph: **A.** Anamorph (*Aspergillus hollandicus*). **T.** Teleomorph (*Eurotium amstelodami*), asci and ascospores. **W.** Teleomorph – ascoma with ascomatal wall (arrow). Scale: 10 µm.

2.2.1 Asexual reproduction (Fig. II.2.4a–e)

Asexual reproduction allows fast multiplication and spreading of an organism. Spores are produced exclusively by mitosis ("mitospores", "mitotic spores") either endogenously or exogenously.

Fruiting bodies

- **Conidiomata** (sing. Conidioma) are structures in or on which non-motile asexual spores (conidia) are contained. **Pycnidia** (Fig. VIII.9d) are closed, often globose or flask-shaped structures lined inside with conidiogenous (conidia forming) cells, whereas **Acervuli** (Fig. VIII.9a, c, j) are plate-like stromatic layers on which conidiophores (conidia-bearing structures) are seated.

- **Synnemata** (sing. **synnema**; Fig. VIII.15d) are bundles of erect conidiophores, on which conidia are formed mainly apically.

Asexual spores (mitospores)

- **Zoospores** are motile, flagellate spores (planospores), found only in fungus-like Protista. This group is not discussed further here.

- **Conidia** are non-motile spores found predominantly in Zygomycota, Ascomycota and Basidiomycota, as well as in some Oomycota (Chromista).

Fungus-like Protista reproduce usually by endogenous zoospore formation and rarely by forming aplanospores (non-motile propagules). Asexual reproduction in the Zygomycota, Ascomycota, and Basidiomycota, on the other hand, is quite diverse:

- In the **Zygomycota,** spores (sporangiospores) are formed endogenously in sporangia (sing. sporangium; e.g. *Mucor fuscus*, *Rhizopus* spp., Zygomycota). Sporangia are swellings formed at the tip of differentiated hyphae, called sporangiophores, in which sporangiospores are produced by mitosis and are released by breaking of the sporangial wall (Figs. II.2.4b–c, VIII.1–4). Sometimes the hypha forming the sporangium may differentiate into a sterile, dome-like structure within the sporangium called **columella**.

- In the **Ascomycota** and **Basidiomycota**, conidiogenous (conidia-producing) cells are formed either directly on the hyphae or on more or less differentiated, erect conidiophores (conidiogenous cells bearing structures). The conidia are formed **exogenously** by the conidiogenous cells and released at maturity (Figs. II.2.4d–e, VIII.8–18). Several forms of conidiogenesis are known and are described in detail in Section VII. For example, the *Penicillium* anamorph of *Talaromyces* (Ascomycota) is characterised by a conidiophore bearing at its tip a brush-like group of conidiogenous cells from which conidia are formed in long chains.

2.2.2 Sexual reproduction (Fig. II.2.4f–k)

Sexual reproduction is fundamental for the genetic exchange and evolution within a given taxon. Approximately 70% of all known fungi and fungus-like Protista reproduce sexually.

Fruiting bodies

- **Zygomycota** form characteristic **zygosporangia** (sing. zygosporangium), which contain only one sexual spore (Zygospore) each (Fig. VIII.3e–g). Under the microscope however, the zygospore is normally not directly visible because of the thick zygosporangial walls.

- Sexual reproduction of **Ascomycota** normally takes place in typical **ascomata** (sing. ascoma; Fig. VIII.5–7); correspondingly, **Basidiomycota** produce their sexual structures in or on **basidiomata** (sing. basidioma). Ascoma and basidioma shape and their overall morphology are crucial for the taxonomic classification and identification of both groups.

Sexual spores (Meiospores)

- In the **Zygomycota, zygospores** are formed endogenously in a **zygosporangium** (e.g. *Zygorhynchus moelleri*, Fig. VIII.3e–g).
- **Ascomycota** produce their **ascospores** endogenously in an **ascus** (e.g. *Sordaria fimicola,* Fig. VIII.7j).
- **Basidiomycota** (e.g. *Boletus edulis*) form exogenously **basidiospores** on **basidia** (Fig. VIII.8b, c).

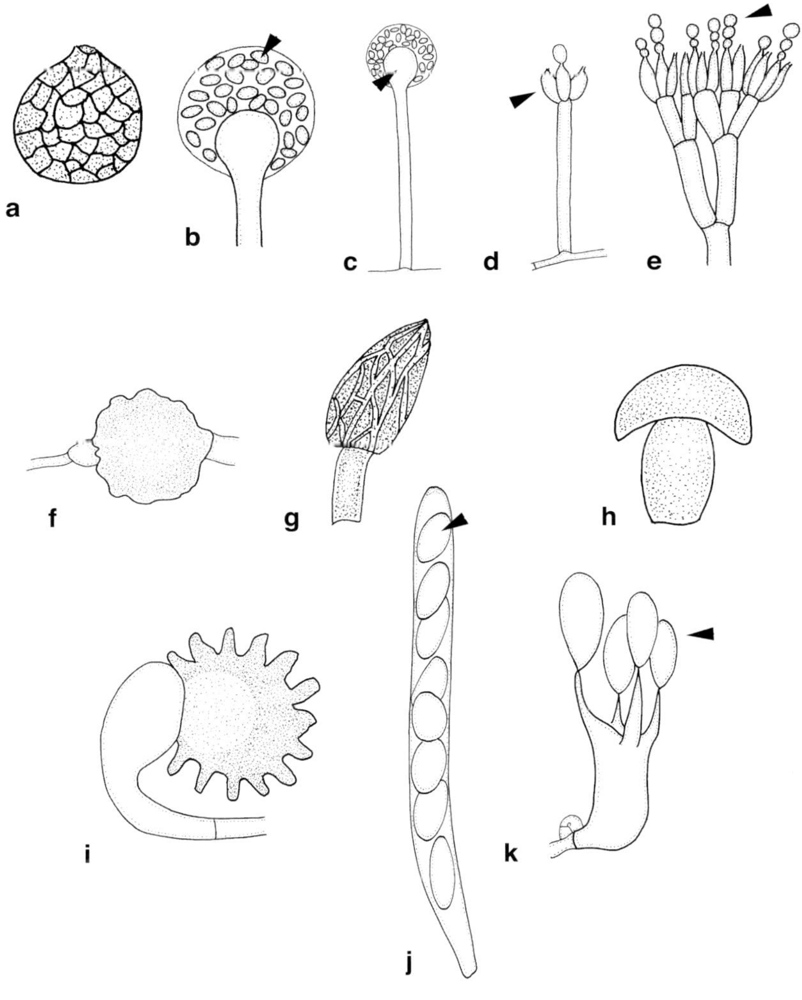

Fig. II.2.4. Reproduction in fungi

a–e. Asexual organs: **a.** Conidioma (Pycnidium). **b.** Sporangium with sporangiospores (arrow). **c.** Sporangiophore with columella (arrow). **d.** Conidiophore with conidiogenous cells (arrow). **e.** Complexly branched *Penicillium* conidiophore with conidia (arrow).

f–k. Sexual organs: f–h. Fruiting bodies: **f.** Zygosporangium. **g.** Ascoma. **h.** Basidioma. **i–k.** Meiospores: **i.** Zygospore in zygosporangium. **j.** Ascospore (arrow) in ascus (endogenous spore formation). **k.** Basidiospore (arrow) on basidium (exogenous spore formation).

III. Taxonomy

1. Phylogenetics, systematics, classification, taxonomy

There is some confusion about the exact meaning of the terms "phylogenetics", "systematics", "classification", and "taxonomy".

Phylogenetics is the study of evolutionary relationships among groups of organisms (e.g. species, populations); thus, it tries to establish the "tree of life". This is done mainly by molecular biology, most notably using sequencing data.

Systematics, on the other hand, deals with the diversity of life: it studies and establishes relationships between organisms in order to classify them according to their phylogeny. Phylogenetics and systematics are only apparently identical: in fact, phylogenetics adds the evolutionary time factor to the study of systematics. The field of **phylogenetic systematics** deals with identifying and understanding the evolutionary relationships among living (extant) and dead (extinct) organisms and combines methods of phylogenetics and systematics.

Classification tries to organise items (in biology: organisms) into well-characterisable groups, while **taxonomy** is the science that tries to describe, name and order organisms.

Taxonomy may go hand in hand with classification, but it needs not: it may include some aspects of phylogenetics, but most often it does not. Its primary purpose is to provide means for the identification of organisms.

2. The 7 kingdoms and the „new“ fungal taxonomy

As all other living organisms, fungi are named according to the binomial system, which describes an organism by a genus and a species name.

Until 1980 fungi were considered a quite heterogeneous but compact taxonomic group. This changed when Cavalier-Smith (1981) described the kingdom fungi as comprising the Ascomycota, Basidiomycota, Zygomycota and their anamorphic forms, as well as the Chytridiomycota. He included fungus-like organisms producing motile spores partly in the Protista and partly in the Chromista. Kendrick (1992) provided a comprehensive representation of this scheme, now accessible also over the Internet (http://www.mycolog.com/fifthtoc.html).

Cavalier-Smith subsequently refined and completed his scheme of the tree of life (Cavalier-Smith 1993, 1998, 2004, 2006). Fungal phylogenetics has been formally described in detail by Hibbet et al. (2007). Presently living organisms are divided in 7 kingdoms (Eubacteria, Archaebacteria, Archaezoa, Protozoa, Plantae, Animalia, Fungi); an additional kingdom, the Chromista, is still controversial.

Presently fungi are classified according to the scheme presented in Fig. III.1.

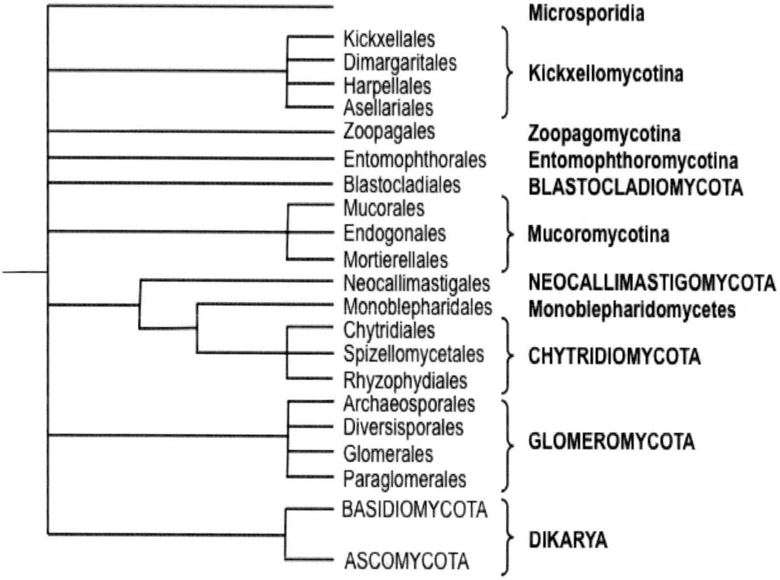

Fig. III.1. Classification of Fungi (slightly modified from Hibbet et al. 2007).

The kingdom Fungi includes the Ascomycota and Basidiomycota (Dikarya), the Glomeromycota (endogenous mycorrhiza), Chytridiomycota, Neocallimastigomycota, Mucoromycotina (Zygomycota), Blastocladiomycota, Entomophthoromycotina, Zoopagomycotina, Kickxellomycotina, and Microsporidia (unicellular human pathogens).

The slime moulds (Myxomycota) are now part of the Protista (Protozoa: Amoebozoa, Eumycetozoa), whereas the Oomycota (water moulds and downy mildews) and the Thraustochytridae are currently placed in the kingdom Chromista.

This classification has been made possible by the progress in molecular biology that has allowed easy and cost-effective sequencing of many organisms. One cannot but notice, however, the excellent correspondence of the new system with the hypotheses that Müller & Löffler (1992) put forward based on morphological and physiological observations.

Morphology and physiology, however, can lead to wrong phylogenetic conclusions, on the one hand because of the pleomorphic nature of fungi, on the other because of the difficulty of interpreting morphological homologies and analogies. Firm phylogenetic conclusions, therefore, can be reached only after careful molecular genetic analysis of all organisms.

Phylogenetic studies of fungi are currently ongoing and lead to continuous updates of the phylogenetic system. The most recent changes in fungal phylogeny and classification may be found in the web pages of the "Tree of Life Project" (http://tolweb.org/tree/).

3. Protista and Chromista

These two groups include a large number of organisms. Fungus-like Protista include the Myxomycota (slime moulds), the Acrasiomycota (cellular slime moulds), the Plasmodiophoromycota (endoparasitic slime moulds), and the Labyrinthulomycota (slime nets – a group of marine saprobes). The Oomycota and the Hyphochytriomycota belong to the Chromista. Of these, only the Oomycota are briefly discussed here.

3.1 Oomycota

Thallus: diploid, coenocytic (aseptate) mycelium.

Digestion: exogenous.

Cell wall: contains cellulose (rarely chitin); synthesis of lysine via the diaminopimelic acid pathway (similar to bacteria) and not via the aminoadipinic acid as all fungi.

Anamorph: zoosporangia producing zoospores with anteriorly or laterally inserted flagella (1 whiplash and 1 hairy flagellum) or conidia formation.

Teleomorph: oospores (oogamy, followed by the formation of oogonia containing oospores).

The Oomycota comprise approximately 600 species, many of which are important crop pathogens. One notable example is *Phytophthora infestans*, an oomycete that may cause an important disease called late blight or potato blight. *P. infestans* was the cause of the Irish and Highland potato famine (1845–1849), which led to the emigration of a large number of Irish to the USA.

Some additional examples are presented in Tab. III.3.1.

Tab. III.3.1: Some economically important Oomycota.

Saprolegnia parasitica Coker	Saprolegniosis, characterized by white or grey patches of mycelium on the body or fins of freshwater fish.
Phytophthora infestans (Mont.) de Bary	Late blight or potato blight. It causes serious losses of potato crops worldwide; probably the most important pathogen of potato and tomato.
Plasmopara viticola (Berk. & Curt.) Berl. & de Toni	Grapevine downy mildew.*
Pythium debaryanum Hesse	"Damping off" disease of seedlings; very common in field and greenhouses, killing newly emerged seedlings.

*) not to be confused with the powdery mildew, which is caused by some Ascomycota (Erysiphales)

4. Fungi

Approximately 70,000 fungal species have been described, although their total number is estimated to be as high as 1.5 millions (Hawksworth 2006). Fungal biodiversity is very high and difficult not only to quantify but also to study (Crous et al. 2009). Here we only list those groups that are important as moulds or pathogens of plants and animals.

4.1 Zygomycota

(Approx. 700 known species)

4.1.1 Zygomycetes (Fig. VIII.1–4)

Thallus: haploid, usually aseptate or (rarely) septate mycelium.

Septal pores: micropores, central pores.

Cell wall: with chitin, glucans and mannan; occasionally chitosan.

Anamorph: sporangia with endogenously formed sporangiospores; occasionally also conidia are formed.

Teleomorph: zygosporangia with one zygospore.

Some examples are presented in Tab. III.4.1.

Tab. III.4.1: Examples of Zygomycota

Mucor mucedo L. : Fr. *Mucor plumbeus* Bon.	On horse manure. On soil, hay, dung.
Rhizopus oryzae Went & Prinsen Geerlings	On dead plants and in soil; lactic acid fermentation.
Absidia spp.	In soil or on meat.
Lichtheimia spp.	Human pathogen, causing zygomycoses.
Entomophthora spp.	Insect pathogens.
Endogone spp.	Vesicular-arbuscular (VA) mycorrhiza (now more correctly placed in the Glomeromycota).

Most Zygomycota are saprobes, but some *Mucor*, *Rhizomucor*, *Rhizopus* and *Lichtheimia* species may cause fungal infections (zygomycoses) in animals and humans (Pfaller & Diekema 2010). All entomophthorales, to which *Entomophthora* belongs, are pathogens of insects.

4.2 Ascomycota (Fig. VIII.5–18)

(Approximately 32,500 species known)

The phylum Ascomycota is quite heterogeneous and species-rich. Sexual reproduction in the Ascomycota is characterised by the formation of asci and ascospores. The group includes taxa of alimentary, pharmaceutical and chemical importance (*Claviceps* spp., yeasts), human (e.g. *Arthroderma* spp.) and plant pathogens [e.g. *Taphrina deformans* (peach leaf curl disease), *Erysiphe* spp. (powdery mildew), *Venturia inaequalis* (Apple scab disease)].

Thallus: haploid, very rarely dikaryotic, septate mycelium with central, simple septal pores.

Cell wall: with chitin, glucan and mannan (very rarely cellulose – cf. *Ophiostomataceae*).

Anamorph: conidia.

Teleomorph: fertilisation occurs after fusion of differentiated cells. This produces a dikaryon: subsequently the two nuclei fuse to produce asci, in which karyogamy occurs, followed by a meiosis (and in most cases a mitosis thereafter, to produce 8 ascospores). The asci are usually contained in an ascoma. Asci, ascospore and ascoma shapes and sizes are relevant taxonomic characters.

Some examples are presented in Tab. III.4.2.

Tab. III.4.2: Examples of Ascomycota

Yeasts	In or on fruit and fruit juices; alcohol fermentation and leavening agent (baker yeast – *Saccharomyces cerevisiae*).
Taphrina deformans (Berk.) Tul.	Peach curl disease.
Blumeria (*Erysiphe*) *graminis* DC	Powdery mildew on cereals.
Uncinula necator (Schw.) Burr.	Powdery mildew on grapevine.
Morchella esculenta (L.) Pers.	Morels.
Eupenicillium spp., *Talaromyces* spp. (Anamorph: *Penicillium*)	Production of penicillin, cheese ripening.
Claviceps spp.	Ergot (holy fire).
Venturia inaequalis (Cke.) Wint.	Apple scab disease.

4.3 Basidiomycota

(Approximately 14,000 known species)

The Basidiomycota comprise approximately 30% of all known fungi. The sexual reproduction is characterised by the formation of basidia, on which basidiospores are produced exogenously. Several species are known to live exclusively as mycorrhiza of trees and shrubs. Many basidiomycetes can be grown in pure culture, but only a few (e.g. *Agaricus bisporus* – button mushroom; *Pleurotus ostreatus* – oyster mushroom) form their fruiting bodies in culture.

Morphologically two groups can be distinguished.

4.3.1 Ustilagomycotina (smuts)

Thallus: budding cells or septate mycelium.

Septal pores: simple central pores (Fig. II.2.1.c).

Teleomorph: zygogamy (fusion of two haploid cells), development of a dikaryon, formation of a probasidium and a promycelium with exogenous sporidia.

4.3.2 Pucciniomycotina (rusts) and Agaricomycotina (mushrooms)

Thallus: haploid or dikaryotic mycelium, with or without clamp connections.

Septal pores: dolipores (Fig. II.2.1.d).

Anamorph: conidia.

Teleomorph: fertilisation of the female cell by a male cell (both morphologically undifferentiated). Formation of a dikaryon, from which the typical fruiting bodies arise. In or on the fruiting bodies: basidia and basidiospore formation (Fig. VIII.8.b, c).

Some examples of Basidiomycota are presented in Tab. III.4.3.

Tab. III.4.3: Examples of Basidiomycota

Smuts	*Ustilago tritici* (Pers.) Rostr. (loose smut of wheat) *Ustilago maydis* (DC) Corda (corn smut)
Rusts	*Puccinia graminis* Pers. (wheat stem rust) *Puccinia glumarum* (Schm.) Eriks. & Henn. (yellow rust) *Uromyces appendiculatus* (Pers.) Lk. (common bean rust)
Mushrooms	*Amanita muscaria* (L.) Lam. (fly agaric) *Boletus edulis* Bull. : Fr. (cep, penny bun, porcino) *Agaricus bisporus* (Lge.) Imbach (button mushroom)

4.4 Anamorphic fungi (mitotic fungi; Fig. VIII.8–18)

Anamorphic fungi (previously known as "Deuteromycetes" or "Fungi imperfecti") are organisms for which the teleomorphic form is not known or has so far never been observed (thus the name "fungus imperfectus" – imperfect fungus).

Most likely many of these organisms are no longer able to reproduce sexually. We include in the discussion (and in the keys) of this group, however, also ascomycetous and basidiomycetous anamorphs with known teleomorphs: on the one hand, many fungi are still mentioned in the literature by either the anamorphic or the teleomorphic name; on the other, many fungi produce in pure culture only the anamorph – therefore, the mycologist dealing with pure cultures will often be confronted mainly with anamorphic forms.

Some examples are presented in Tab. III.4.4.

Tab. III.4.4: Examples of fungal anamorphs

Alternaria alternata (Fr.) Keissler	Leaf spots on more than 380 host species.
Alternaria solani (Ell. & Mart.) Sor.	(early) Tomato blight.
Botrytis cinerea Pers.	Grey mould on grapes and other fruits.
Colletotrichum spp.	Anthracnose of several plants.

5. Polyphasic Taxonomy

Fungal taxonomy has so far relied heavily on morphology. Formerly, only in rare cases taxonomists have used physiological and biochemical analyses to characterise fungi. The study of pure cultures in fungal taxonomy has then allowed mycologists to further characterise fungal taxa using biochemical and physiological tests. Initially taxonomists investigated specific pigments and other secondary metabolites as potential taxonomic characteristics (Whalley & Greenhalgh 1971; Whalley & Whalley 1977; Whalley & Edwards 1995), later on biochemical and molecular biology studies were carried out to study phylogenetic relationships among fungal taxa (Berbee & Taylor 1995; Bruns et al. 1991; Eriksson 1995; Kohn 1992; Lutzoni & Vilgalys 1995; McLaughlin et al. 1995; Samuels & Seifert 1995; Stadler et al. 2001a, 2001b; Theler 1995). Molecular biology and proteomic tools have now been established firmly in the mycology lab, without, however, having replaced classical morphological investigations.

So far, fungal taxonomists have rarely attempted to combine the results of different methods into one comprehensive analysis. Mostly, they limited themselves to compare the results of molecular studies with those of morphological and biochemical ones: numerical analyses combining morphological, biochemical and molecular biology data are still quite rare.

In mycology, steps in this direction were first taken by Petrini et al. (1989) and Sieber-Canavesi et al. (1991) with two conifer-inhabiting ascomycetous genera, and later on by Sieber et al. (1991) with some alnicolous *Melanconium* species. These are examples of what is currently called "polyphasic taxonomy", a term coined by Colwell (1969, 1970) to define her approach to the taxonomy of *Vibrio cholerae*. Polyphasic taxonomy (which should perhaps be better called "integrated taxonomy") relies on combined analyses of all morphological, biochemical, ecological, geographic, molecular, and proteomic data produced in a taxonomic study of a given group of organisms (Petrini & Petrini 1996).

This approach enhances understanding of the taxonomic position of a taxon by providing a broader, more comprehensive analysis of all available characteristics. The downside is the need of an integrated approach by an interdisciplinary team of taxonomists, molecular biologists, ecologists and numerical taxonomists. The right choice of the appropriate multivariate numerical methods is admittedly difficult and more time is needed to complete such a study, but the quality of the results clearly outweighs these drawbacks.

Any method may be used in polyphasic taxonomy, provided it is sensitive, specific and accurate enough to characterise a fungal taxon. Light and fluorescence microscopy, as well as substrate utilisation tests have been successfully used in a number of cases (e.g. Petrini et al. 1989; Sieber-Canavesi et al. 1991). Ecological data are also helpful, because they provide important information on the adaptation of an organism to its ecosystem (e.g. L. Petrini et al. 1987). When molecular biology methods were not yet as widespread and cost-effective as they now are, isoenzyme analyses (e.g. Brunner & Petrini 1992; Leuchtmann et al. 1992) as well as the study of secondary metabolites (Frisvad & Filtenborg 1989; Svendsen & Frisvad 1994) were extensively used to taxonomically characterise fungal taxa. Presently, however, they are increasingly replaced by molecular biology (RFLP, Sequencing) and proteomic (MALDI-TOF mass spectrometry, see Section IV.3) methods, because they are easier to use, more cost-effective and reliable than the more cumbersome biochemical approaches. The computing power available today, combined with user-friendly and powerful statistical software allows rapid processing of the data produced during a polyphasic taxonomic study. This has changed quite substantially since the time numerical taxonomy was first introduced (Sneath 1989, 1995).

Data analysis should be carefully planned and follow a step-wise approach. Most taxonomists immediately jump to the use of confirmatory statistics, often forgetting the advantages offered by a thorough exploratory, descriptive inspection of the data by exploratory data analysis (EDA) techniques (Cleveland 1994; Tukey 1977). These methods (e.g. boxplots, confidence intervals and confidence ellipses), routinely available in most statistical data analysis software packages, allow a rapid and robust inspection of the data that is often sufficient to draw reliable conclusions on the taxonomic grouping of taxa.

Numerical methods currently used in polyphasic taxonomic evaluations include techniques discussed by Sneath & Sokal (1973) in their seminal work on numerical taxonomy. Classical clustering algorithms, block and fuzzy clustering, principal components, principal coordinate analysis, factor analysis, and non-metric multidimensional scaling (MDS) have been extensively described as valid options for numerical taxonomy by several authors (Ludwig & Reynolds 1988; Pankhurst 1991; Sneath & Sokal 1973) and are offered by most statistical packages.

Multiple correspondence analysis (Greenacre 1986, 1989, 1993; Petrini & Sieber 2000; Sieber et al. 1998) has often been applied to marketing and ecological data, but is still rarely considered for taxonomic studies. Although it may be used for confirmatory purposes, it is more of an exploratory technique that lends itself to analysing taxonomic data. It is quite robust when data are missing (incomplete) but it is sensitive to outliers.

Phylogenetic analysis packages should not be routinely used with phenetic data, even if in some cases it has been done with some success (L. Petrini 1992). Its application to mixed morphological and molecular data is still controversial and only few examples of successful interpretations are available (e.g. Lutzoni & Vilgalys 1995).

IV. The identification of moulds

1. General considerations

Moulds are fungi that grow usually saprobically on organic substrates and may cause modifications of the substrate's appearance, texture, and sometimes (in the case of food) taste. "Mould" is no taxonomic physiological or biochemical term. Moulds are fungi, but not all fungi should be called moulds. Systematically, moulds belong to the Mucorales (Zygomycota), Ascomycota and Basidiomycota (both teleomorphic and anamorphic forms), as well as to the Oomycota (e.g. *Peronospora tabacina* – the blue mould of tobacco, causing bluish-grey mildew on the leaf underside).

The term is not linked to the practical significance of the organism either. Moulds can cause disease or food spoilage, some play an important role in biodegradation, others in the production of food (e.g. *Penicillium roqueforti* in blue cheese), antibiotics (*Penicillium* spp. as penicillin producers), enzymes (cellulases produced by *Trichoderma* spp.), or organic substances (citric acid by *Aspergillus niger*), whereas some may be opportunistic human and animal pathogens.

This guide is intended to be an introduction to the identification of the most common and important genera of moulds. It considers only those genera that fruit readily in pure culture and are commonly found in the environment and in food, are common plant pathogens or may be isolated from human and animal tissues as opportunistic pathogens.

2. Morphological characteristics

Moulds may be identified by morphological characteristics to the genus and sometimes to the species level only if fruiting organs are formed. Their appearance in pure culture and their growth rate on selected media are important identification features.

3. Chemotaxonomic and molecular biology techniques

Morphology is a good start to identify moulds, but it is often insufficient to achieve a reliable identification to the species level. For instance, species in the anamorphic genera *Aspergillus*, *Fusarium* and *Penicillium* are hardly distinguishable by morphology alone, and then often only by the experienced eye of the expert. Molecular biological methods and, to a lesser extent, chemical and biochemical characteristics are now used routinely in the lab to identify fungi (Frisvad et al. 1998; Samson & Pitt 2000).

For the identification of many food-borne fungi chemotaxonomy still plays an important role. Secondary metabolic products, determined by thin-layer chromatography or HPLC (Frisvad & Thrane 1987; Frisvad et al. 1989; Samson & Pitt 2000) can be quite species-specific and help in the identification of *Aspergillus*, *Fusarium* and *Penicillium*. While chemical and biochemical methods are also very useful to check for species- or strain-specific production of mycotoxins, they are often time-consuming and relatively expensive: furthermore, their use is not always straightforward and they are often reserved, therefore, to specialised labs.

Immunological techniques have been in use for a long time for the identification of food-borne and human pathogenic fungi (De Ruiter et al. 1993; Bille & Schär 2001). Some commercial kits are available also for the identification of plant pathogenic fungal species, but molecular biological methods are now the first choice in mycological taxonomy studies. Techniques such as Restriction Fragment Length Polymorphism (RFLP), Random Amplified Polymorphic DNA (RAPDs: e.g. Williams et al. 1990; Hämmerli et al. 1992), Polymerase Chain Reaction (PCR), and sequencing, to name a few, are increasingly used in the mycological labs, because they are comparatively fast, reliable, and cost-effective. Sequencing of the ITS-1 and ITS-2 regions has established itself as a very effective tool in fungal taxonomy and to-date a large number of reference sequences are available on the Internet (see for example „Assembling the Fungal Tree of Life Project“; http://www.aftol.org/index.php). The ITS sequence data are considered extremely reliable for the identification of a large number of fungi and are now routinely used in many fungal barcoding projects (http://www.allfungi.com/index.php). DNA barcoding, a taxonomic method that uses short genetic markers to characterise a particular species, has as the main goal to identify an unknown sample in terms of a known classification (http://www.barcodeoflife.org/). Other genes used for barcoding fungi include *cox1* (cytochrome oxidase I) in *Penicillium* (Seifert et al. 2007) or beta-tubulin in *Aspergillus* (Balajee et al. 2007). For some fungal taxa extensive databases of molecular data are available, to which sequences can be submitted, allowing identification of the sequenced specimen by sequence comparison (e.g. „TrichoKEY: http://www.isth.info/tools/molkey/index.php?do=unset; Druzhinina et al. 2005).

Fluorescence *in situ* hybridisation (FISH) has not yet found a prominent place in the mycological routine lab. Some kits, however, are now available on the market for the direct (partial) identification of human pathogenic *Candida* species in blood samples (e.g. Forrest et al. 2006).

Proteomics is now finding its place in mycology as well, after having been used successfully in the identification of bacteria. MALDI-TOF MS (matrix-assisted laser desorption ionisation – time of flight mass spectrometry) is a powerful method to identify organisms based on their protein spectra (Wilkins & Lay 2006). MALDI-TOF MS has proven to be very reliable for the identification of bacteria (Seng et al. 2009), yeasts (Marklein et al. 2009) and more recently filamentous fungi and it has been successfully applied, for example, to the taxonomy of the ascomycetous genus *Hypocrea* and its anamorphs *Trichoderma* (De Respinis et al. 2010; Samuels et al. 2010). The dendrogram resulting from the cluster analysis of MALDI-TOF MS data was roughly identical to the classification obtained from ITS and *tef1* sequences. While the costs of a MALDI-TOF mass spectrometer are relatively high, those of the consumables are negligible – the technique is easy to use and the identification is fast and reliable. While it will not replace morphology and molecular biology in serious taxonomic work, MALDI-TOF is most likely to become the method of choice for routine identifications in mycology.

V. Working instructions for the identification of moulds in pure culture

1. General comments

Fungi are generally identified based on their macro- and micromorphological characteristics. An accurate identification is only possible if the morphological features of the organism in question have been properly characterised. Macromorphology and appearance in culture (colour, colony growth rate) are important features and must be noted. A reliable identification, however, is possible only after careful observation of microscopic characteristics.

1.1 Observation of morphological features

Only rarely can a mould be identified directly by looking at its macroscopic characteristics and cultural aspects. A conclusive identification by morphology is only possible after observing and studying its vegetative and/or sexual reproductive structures. As a rule of thumb, a fungus can be safely identified

if spores are seen

and

if the mode of formation of these spores can be observed.

To avoid jumping to wrong conclusions, however, care should be taken to examine cultures

- not too young, because the relevant features may not yet be present or only rudimentary;
- not too old, because lysis may have already set in and the shape and size of the reproductive organs (in particular spores) is not representative – in addition, in many cases the genesis of the spores may no longer be observed.

Moulds in culture do not always sporulate readily and cultures may remain sterile for a long time, sometimes weeks. Cultures still sterile after 4 or 5 days must not necessarily remain sterile for good. Changing culture conditions (e.g. exposing them to UV light, transferring them to special media (see Section VI.1) may often help. It is not uncommon to see a fungus fruiting in pure culture after 6 or more weeks!

Polymorphism (formation of more than one type of fructification) in fungi is not uncommon – quite often, different reproductive forms are observed in cultures of different age or on different media. It may be advisable, therefore, to study cultures at several time points, or even to inoculate the fungus to be identified on different media. As a rule, this is not often required, but it may help in difficult cases.

Resorting to cultural work should be the rule. Isolating a fungus in pure culture is usually done by directly streaking some of the fungal material on a nutrient agar, by dilution series or by baiting. These and other methods are described in several textbooks, e.g. Johnston & Booth (1983). Isolation in pure culture is almost always a crucial step for the successful identification of a mould and for any subsequent work (biochemical analyses, molecular biology investigations). Selective media for fungi with special ecological and physiological requirements (xerophile, heat resistant species) are also routinely used (Samson et al. 1992, 2004).

1.2 Identification by means of (dichotomous or synoptic) keys

Identification of a fungus is principally carried out using dichotomous or synoptic keys. For some moulds, computerised keys are now available. The use of all types of keys, however, requires in-depth knowledge of technical terms, of which there are many, not necessarily easy to retain. We describe and illustrate the most important ones in sections VII on page 35 and VIII on page 49. Additional information can also be found in section X on page 137.

1.3 Safe handling of moulds

Most moulds can be handled without any strict safety precautions. Some, however, may represent a safety hazard for animals and humans. It is therefore advisable to treat all of them as if they were potentially hazardous to the operator's health.

Some moulds do not only form spores but also release toxic volatile compounds (VOC: "volatile organic compounds"). Spores and VOC, if inhaled, may cause allergic reactions, asthma or allergic (hay) fever in immunocompromised or immmunocompetent, predisposed hosts (Mücke & Lemmen 2004).

Most moulds are classified as belonging to biosafety level (BSL) 1 and 2 and only a handful of species to BSL 3 [BSL 1: not known to consistently cause disease in healthy adult humans, and of minimal potential hazard to laboratory personnel and the environment; BSL 2: agents of moderate potential hazard to personnel and the environment, that cause only mild disease to humans or are difficult to contract via aerosol in a lab setting; BSL 3: agents that may cause serious or potentially lethal disease after inhalation but for which treatments exist (CDC 2009)]. One should not forget, however, that these definitions are theoretical: the line between pathogenic and non-pathogenic organisms is very thin and often an individual's predisposition is relevant for the development of a disease. Immunocompromised individuals, in general, are more likely to be colonised or even infected by fungi. In any case, both immunocompromised and immunocompetent people should handle clinical fungal specimens and media with utmost care.

The use of aseptic and sterile techniques, good personal hygiene, avoidance of bioaerosol inhalation during the work and careful sterilisation of all tools before and after work should be fostered in all laboratories. Fungal cultures should not be opened unnecessarily; "culture sniffing" (Barkham & Taylor 2002) should be avoided especially if medically important organisms are suspected to be growing in a dish. Loops and needles used for inoculating substrates or preparing microscopic slides must be cooled before immersing them in the fungal colonies to avoid the formation of bio-aerosols. We recommend using lactic acid instead of lactophenol (carcinogenic) for microscopy. If lactophenol is used, inhaling its fumes should be avoided.

2. Microscopy

2.1 Direct microscopy

Direct microscopy of mould lawns on the substrate is helpful only when conspicuous structures are observed macroscopically and vegetative or sexual spore formation is suspected. This may lead to a rapid identification, but we recommend also in these cases to verify the identification by isolating the mould in pure culture.

2.2 Pure cultures

Sporulating structures of many fungi have been described from pure cultures. This is the reason why we recommend careful check of results from direct microscopy by comparison with data derived from pure cultures, if possible from colonies grown on the nutrient media used by the authors of the respective taxonomic monographs. The most popular media are briefly described in section VI on page 31.

Liquid culture media are not particularly useful to study fungal structures. Culturing should be carried out in Petri dishes containing 20–25 cm^3 of solid medium, incubated at room temperature (ideally at two different temperatures, e.g. 20 and 27 °C) and 37 °C, to allow for detection of thermotolerant and thermophilic fungi: for clinical specimens, incubation at 37 °C is mandatory, but we recommend to store some plates at 24–27 °C as well, because some opportunistic pathogens will grow more rapidly under these conditions. If sporulation does not occur spontaneously within 10 days, incubation under near UV light (256 nm) or in daylight may help.

Microscopic examination of sporulating cultures will usually include:

- **Checking the intact thallus under a dissecting microscope.** If the culture has no powdery appearance (a powdery culture suggests extensive spore formation, with potential distribution of spores in the air), one could consider cutting out small blocks of the culture, putting them on a slide and observing them directly (without any cover slip) under the lowest microscope magnification, being careful not to touch the material with the objective. In some cases, this allows direct observation of conidial chains.
- **Studying the fungal structures under a light (or phase contrast) microscope.** Care must be taken to prepare thin preparations with as little nutrient medium (agar) as possible. Mycelia and putative fruiting structures may be extracted from the colonies using a sterile needle or a forceps.

The first slide should always be a water mount, to check for the presence or absence of special structures such as slimy sheaths or appendages, which may dissolve in other mounting or staining solutions such as lactic acid or cotton blue. Lactic acid is then the mounting medium of choice – lactophenol should be avoided, because it does not provide any advantage over lactic acid. If no phase contrast microscope is at hand, cotton blue may be used to stain fungal structures. Fluorescence microscopy (using calcofluor as staining medium) may be appropriate to visualise phialides and anellides in some anamorphic genera. Fluorescence microscopy is also the method of choice for microscopic visualisation of DNA (e.g. DAPI staining; cf. also Streiblová 1988). The most commonly used mounting media are briefly described on page 33.

All microscopic preparations should be covered with a cover slip and the microscopy study should start at the lowest magnification (10×). If lactic acid is used as the mounting medium, the slide may be briefly heated to eliminate air bubbles that could disturb observation. Slides prepared with lactic acid can be stored for a long time (up to 6 months) without them drying out. Adding lactic acid mostly restores the preparation to its original conditions.

The colour (top and reverse), shape and growth rate of the cultures should be recorded as well.

VI. Culture and mounting media

1. Culture media

Several culture media are available for routine mycological work. One of them, malt agar, is probably the easiest to produce and is available as ready-made powder or Petri dishes from a number of manufacturers. Most descriptions of fungi sporulating in pure culture refer to growth on malt agar, but for a reliable identification to the species level it is crucial to use the same medium described in the original monographs of the genus under investigation.

The media listed here alphabetically are those most frequently used in mycology. Recipes have been taken from Booth (1971a) and Samson et al. (2004). Crous et al. (2009) provide a very exhaustive list of media employed for fungal cultivation. Most of these media are commercially available as powders or ready-made plates. This increases standardisation and reproducibility of results, but it may be more expensive.

Creatine agar (CREA, for *Penicillium* species)

3 g creatine; 30 g sucrose; 0.5 g Kcl; 1.3 g $K_2HPO_4.3H_2O$; 0.5 g $MgSO_4.7H_2O$; 0.01 g $FeSO_4.7H_2O$; 0.05 g bromocresol purple; 15 g agar; add dest. H_2O to a final volume of 1,000 ml. Adjust the pH to 8.

Czapek Agar (CZ, for *Penicillium* species)

3 g $NaNO_3$; 1 g K_2HPO_4; 0.5 g $MgSO_4.7H_2O$; 0.5 g KCl; 0.01 g $FeSO_4.7H_2O$; 30 g sucrose; 15 g agar; 1 ml trace elements solution; 1,000 ml dest. H_2O.

Czapek-Yeast Agar (CYA, Czapek-Yeast agar, for *Penicillium* species)

Add 5 g yeast extract to CZ.

Leguminosae Agar

1,000 ml tap water; 2 g agar; 20 g leguminosae meal. A variation of this recipe recommends to first boil 50 g meal in 1,000 ml water, filter the suspension, add water to 1,000 ml and finally 2 g agar.

Malt agar (MEA, Difco or Bacto; ideal for growing most fungi)

1.5–2% malt extract; 2% agar; tap water.

Malt agar with 20% sucrose (M20; used to grow osmophilic *Aspergillus* and *Eurotium* species)

1.5–2% Malt extract; 20% sucrose; 2% agar; tap water.

Oatmeal agar (OA, e.g. for Xylariaceae)

Boil 30 g oatmeal in 1 l water during 2 hours, filter the suspension and add 15 g agar. Add tap water to 1,000 ml.

Potato-Dextrose-Agar (PDA, used in particular for *Fusarium* and *Trichoderma*)

230 ml potato extract, produced by filtration of 300 g potatoes boiled during 1 hour in 900 ml water; 20 g dextrose (glucose); 15 g agar; add dest. H_2O to a final volume of 1,000 ml. Adjust the pH to 6.5–7.

Sabouraud Agar (for dermatophytes)

15 g agar; 10 g peptone; 40 g glucose. Combine all ingredients in approx. 900 ml water, adjust to pH 5.6 and adjust final volume to 1,000 ml.

Special Nutrient-poor Agar (SNA, for *Fusarium*)

1 g KNO_3; 1 g KH_2PO_4; 0.5 g $MgSO_4.7H_2O$; 0.5 g KCl; 0.2 glucose; 0.2 g sucrose; 20 g agar; 1,000 ml dest. H_2O.

Yeast extract sucrose agar (YES, for *Penicillium*)

20 g yeast extract; 150 g sucrose; 0.5 g $MgSO_4.5H_2O$; 20 g agar; 1,000 ml dest. H_2O.

Trace elements solution (in particular for *Penicillium* and *Aspergillus*)

1 g $ZnSO_4.7H_2O$; 0.5 g $CuSO_4.5H_2O$; dilute in 100 ml H_2O.

2. Mounting media

Lactic acid:

concentrated L(+) lactic acid

Lactophenol

Phenol	20 g
Conc. L(+) lactic acid	20 g
Glycerin	40 g
Water	20 g

Cotton blue

concentrated L(+) lactic acid	100 ml
anilin blue	1 g

3. Fluorescence microscopy

3.1 DAPI nuclear staining

Dissolve 1 µg/ml DAPI (4‘,6‘–diamidino–2–phenylindole) in 2 mM Tris-HCl buffer, pH 7.5–8. Using a buffered solution is not absolutely mandatory, but it increases the fluorescence contrast. Mount the fungal material (spores or hyphae), incubate for 15–30 minutes in the dark and observe at UV wavelength. Nuclei and other structures containing DNA fluoresce blue.

3.2 Cell wall staining with calcofluor

Dissolve a 0.1–0.5 µg/ml solution of calcofluor (a.k.a. Fluorescent brightener 28, Tinopal LPW) in 0.1 M Tris-HCl buffer, pH 7.5–8. As for DAPI staining, using a buffered solution is not absolutely mandatory, but it increases the fluorescence contrast. A higher calcofluor concentration may result in too bright background fluorescence. Mount the fungal material (spores or hyphae), incubate for 15–30 minutes in the dark and observe at UV wavelength.

Note: contrary to some notes in the literature and conventional wisdom, calcofluor does not specifically stain fungal cell walls. Polysaccharides, including plant-specific ones, may be stained with calcofluor. According to Streiblová (1988), calcofluor may be used also as a vital stain.

VII. Morphological characters

1. Zygomycota (Fig. VII.1.1–3)

The mycelium is aseptate (coenocytic; Fig. VIII.3h), with relatively broad hyphae (up to 4–6 µm in diameter). Hyphae bearing fruiting structures, however, may have some septa. Cultures are often (but not always!) grey, with abundant aerial mycelium and, with some exceptions, quick-growing.

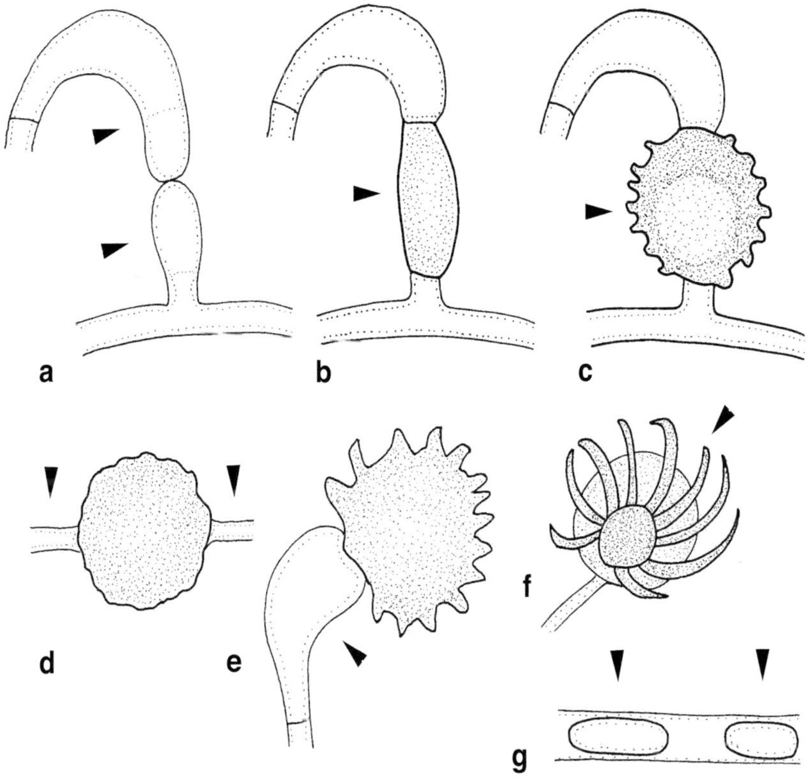

Fig. VII.1.1. Teleomorphs of Mucorales (Zygomycota)

a–c. Zygosporangium formation: **a.** Two physiologically and genetically determined hyphal tips form two zygophores (suspensor cells, arrows). **b.** Gametangia are separated from the zygophores by septa. Formation of a young zygosporangium (arrow). **c.** Thick-walled, mature zygosporangium with ornamented surface (arrow). **d.** Zygosporangium with symmetric zygophores (arrows). **e.** Zygosporangium with asymmetric zygophore (arrow). **f.** Zygosporangium with suspensor adorned with appendages (arrow). **g.** Chlamydospores formed endogenously in a hypha (arrows).

Zygosporangia are conspicuous, dark brown, thick sexual structures (Fig. VII.1.1a–f, VIII.3e–g) that contain one zygospore each. Not all Zygomycota form zygospores in culture. Complex genetic compatibility issues govern their formation.

Sporangiophores (asexual fructification organs) may form in the mycelium on vegetative hyphae or on **stolons** that adhere to the substrate by means of **rhizoids** (holdfasts; Fig. VII.1.2a–c). Sporangiophores may be unbranched or irregularly, dichotomously or sympodially branched (Fig. VII.1.3m–n).

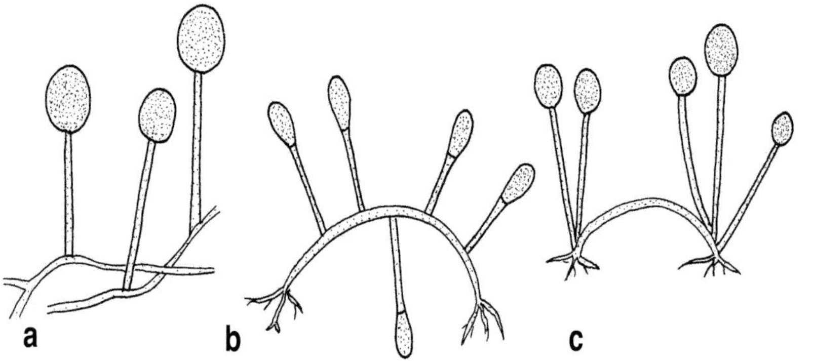

Fig. VII.1.2. Anamorphs of Mucorales (Zygomycota) – 1

a–c. Origin of sporangiophores: **a.** Singly on hyphae. **b.** On stolons. **c.** On rhizoids, linked by stolons.

Sporangiospores are formed endogenously by mitosis (asexually) in sporangia (sing. sporangium) by inclusion of nuclei in plasma portions that are later surrounded by a cell wall. Ripe sporangiospores are then released into the environment by sporangial wall rupture (Fig. VII.1.3a–e, VIII.2a–e).

Depending on genus or species, sporangia may contain a variable number of sporangiospores. The term **sporangium** is conventionally reserved for structures with a large number of sporangiospores (Fig. VII.1.3d, f–g; VIII.1a; VIII.2f), whereas those containing only a few (in exceptional cases one) sporangiospores are termed **sporangioles** (Fig. VII.1.3j–l; VIII.1b–d; VIII.2i–j). **Merosporangia** (sing. merosporangium) are special sporangioles: they are cylindrical and formed radially on a swelling of the sporangiophore (Fig. VII.1.3i; VIII.1e). In the merosporangia, sporangiospores are arranged in a row.

Some sporangia may contain a sterile structure, which forms as a swelling of the distal part of the sporangiophore and is called **columella**. Its function has not been conclusively elucidated but it is generally considered to be of structural support. The columella can be easily seen after rupture of the sporangial wall as a swelling of the sporangiophore. Often remnants of the sporangial wall are attached to it (Figs. VII.1.3c–e, h; VIII.2e; VIII.3a, c; VIII.4e).

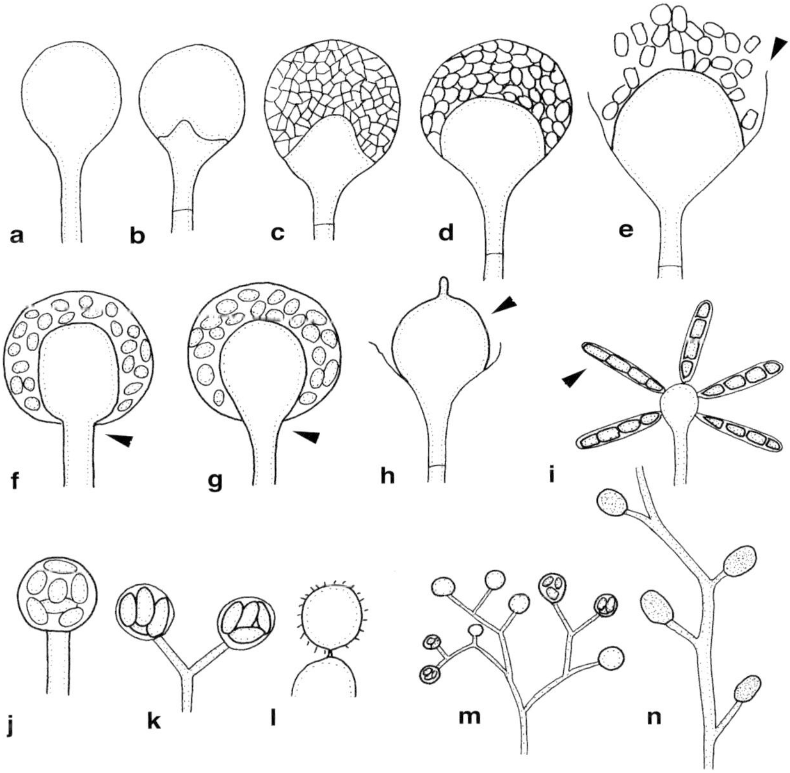

Fig. VII.1.3. Anamorphs of Mucorales (Zygomycota) – 2

a–e. Sporangium formation: **a.** Swelling of the tip of the sporangiophore. **b.** Differentiation of sporangiophore (in some genera by septum formation) and columella. **c.** Columella enlarges in the central part of the sporangium; the cytoplasm migrates to the periphery of the sporangium and encloses nuclei. **d.** Differentiation of sporangiospores by plasma splitting. **e.** Sporangial wall breaks up, sporangiospores are freed; arrow indicates remnants of the sporangial wall on sporangiophore. **f.** Sporangium without apophysis (arrow). **g.** Sporangium with apophysis (arrow). **h.** Columella (arrow) in a sporangium with apophysis. **i.** Merosporangium (arrow) with four spores in a row. **j.** Terminal sporangiole. **k.** Sporangiole on a dichotomously branched sporangiophore. **l.** Sporangiole with only one sporangiospore on a vesicle. **m.** Dichotomously branched sporangiophore with sporangioles. **n.** Sympodially branched sporangiophore.

The **apophysis** (Fig. VII.1.3g; VIII.2d, g, VIII.4e) is a funnel-shaped enlargement of the sporangiophore where the columella begins. This is a genus-specific taxonomic character of the Zygomycota, as is the lack of an apophysis (Fig. VII.1.3f; VIII.3a, c–d).

2. Teleomorphic Ascomycota (Fig. VII.2.1)

The mycelium of the Ascomycota is regularly septate and the hyphae are usually thinner than those of Zygomycota. Asci are distinctive features of the Ascomycota. Only rarely are asci formed directly in the mycelium: most commonly they are enclosed in fruit bodies collectively called **ascomata** (sing. ascoma). Several types of ascomata have been described and the most important are listed here.

- **Cleistothecia** (sing. cleistothecium) are more or less spherical, closed fruit bodies. Their wall (**peridium**) ruptures when asci are ripe, releasing them into the environment. The cleistothecial wall may be composed of more or less tightly interwoven hyphae (they are then called gymnothecia) or may have a cellular structure (Figs. II.3.2w; VII.2.1a; VIII.5g; VIII.6f). Occasionally they may be enclosed in a tight, more or less resistant tissue called **stroma** (pl. stromata).

- **Perithecia** (sing. perithecium) are variously shaped, usually pyriform (pear-shaped) fruit bodies with a distinct opening (**ostiolum**) at the top, through which ascospores ooze and are released. The peridium is usually quite differentiated and comprises one or more layer(s) of cells (Fig. VII.2.1b, VIII.7g). While perithecia formed in culture are mostly scattered on the mycelial surface, on the original host they may be either free or immersed in the substrate. Plant pathogenic Ascomycota, in particular, may form stromata in which several perithecia are contained: contrarily to pseudothecia (see below), in this case, the perithecial wall can usually be distinguished quite easily from the stroma.

- **Pseudothecia** are a particular modification of perithecia: the asci are contained in stromatic, perithecia-like structures: the perithecial wall cannot be distinguished from the stroma and ostiola are not present. This type of fruit body is characteristic of bitunicate taxa.

- **Apothecia** are cup-shaped, shallow fruit bodies, usually seated directly on the substrate or on a differentiated hyphal mat called subiculum. Their peridium is mostly differentiated and multi-layered (Fig. VII.2.1c).

Asci (sing. ascus) normally form in a layer called **hymenium** and may vary in shape. Asci are membranous containers of various shapes in which meiosis takes place. In most cases meiosis is immediately followed by mitosis, producing 8 ascospores. Ascospores are therefore meiospores, as are basidiospores on the basidia (and all sexually formed spores in the fungi in general). The ascal structure is of paramount importance for the classification of ascomycetes.

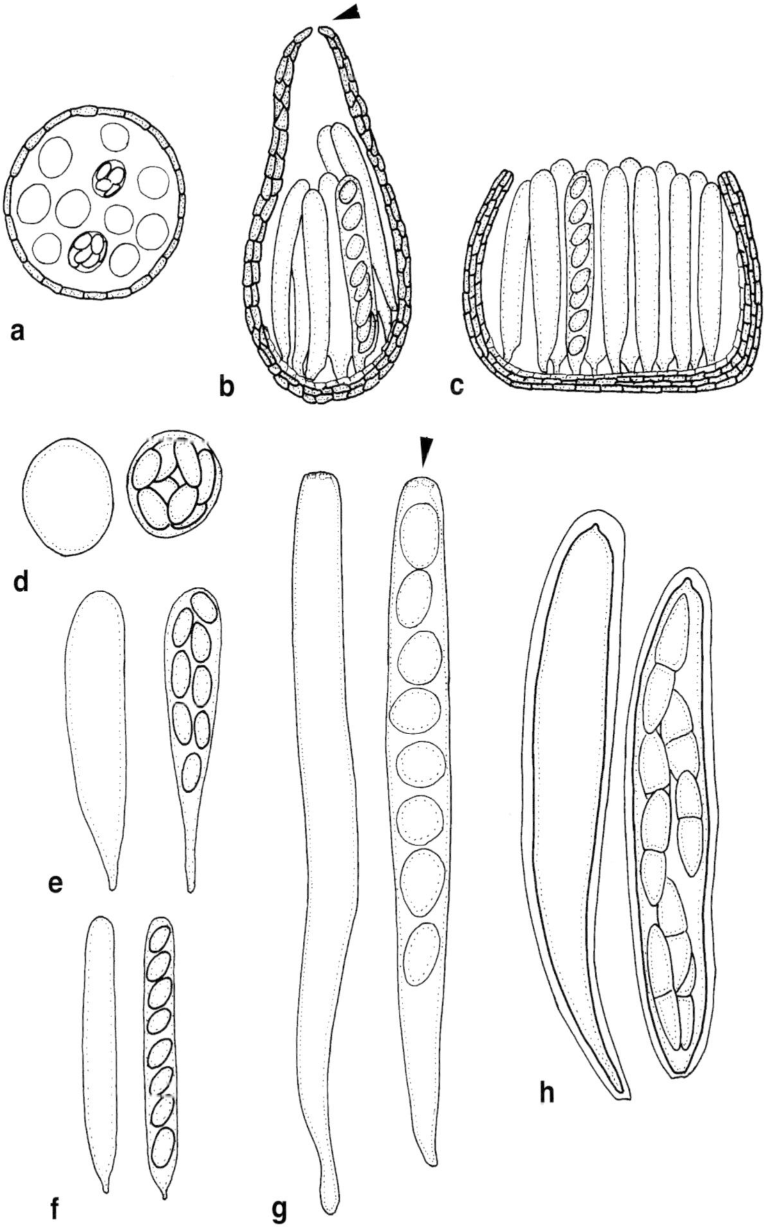

Fig. VII.2.1. Teleomorphic Ascomycota

a–c. Ascomata: **a.** Cleistothecium. **b.** Perithecium with ostiolum (arrow). **c.** Apothecium. **d–h.** Asci: **d.** Spherical, prototunicate asci. **e.** Club-shaped, prototunicate asci. **f.** Cylindrical, prototunicate asci. **g.** Unitunicate asci with apical pore structure (arrow). **h.** Bitunicate asci.

- **Prototunicate asci** are characterised by a very thin ascus membrane that ruptures when the ascospores are ripe. Prototunicate asci are in general spherical, broadly club-shaped or (more rarely) cylindrical (Fig. VII.2.1d–f; VIII.5a, h; VIII.6a, c; VIII.7a, e). They are usually contained in cleistothecia.
- **Unitunicate asci** have a well-developed ascus membrane, which in light microscopy appears as a one-layered structure (Fig. VIII.7j). They are frequently cylindrical or narrowly club-shaped and are mostly formed in perithecia or apothecia. Ascospores are ejected actively from the ascus through an apical pore (Fig. VII.2.1g), which can be variously shaped and sometimes provided by a distinct apical structure ("the ascal plug").
- The wall of the **bitunicate asci** appears two-layered in light microscopy (electron microscopy reveals that the wall has actually more than two layers). The two layers have different elastic properties. The outer layer ruptures readily, whereas the inner one stretches and eventually breaks to free the spores. This is the so-called "jack-in-the-box" ascus type (Fig. VII.2.1h; VIII.8a). Bitunicate asci are usually cylindrical or saccate (shaped like a sac or pouch). Ascospores of bitunicate ascomycetes are mostly two or more celled, with transversal (phragmospores) and longitudinal septa (dictyospores). Bitunicate asci are contained in pseudothecia, embedded singly or to several in more or less conspicuous stromata. Bitunicate ascomycetes are generally plant pathogens or saprobes and are only exceptionally observed in food or as human pathogens.

3. Anamorphic Ascomycota and Basidiomycota (Fig. VII.3.1–3)

The mycelium is regularly septate and conidiogenous cells are formed either directly on the hyphae or on differentiated conidiophores. These are usually septate, conspicuous, branched or unbranched structures that are easily distinguished from the other hyphae (Fig. VII.3.1a–b). They can be formed singly, in bundles (synnemata, sing. synnema; Fig. VII.3.1c), in more or less finite layers (sporodochia, sing. sporodochium; Fig. VIII.14e; acervuli, sing. acervulus), or closed structures (pycnidia, sing. pycnidium; Fig. VII.3.1d). Anamorphic forms of Ascomycota and Basidiomycota are sometimes termed "Hyphomycetes" or "Coelomycetes" in the mycological literature, depending on the type of structures formed. This distinction is not taxonomic and has a purely practical character.

If the conidiophores are formed in more or less closed fruit bodies, the anamorphic forms are commonly called "Coelomycetes". Fully closed [sometimes ostiolate (with an opening, see glossary)] fruit bodies are termed pycnidia, whereas open, cup-shaped, disc-shaped and flat structures are known as acervuli (Fig. VII.3.1d, e). Both can be seated in or on more or less well-developed stromata. In practice, however, distinguishing the two forms in culture is often very hard.

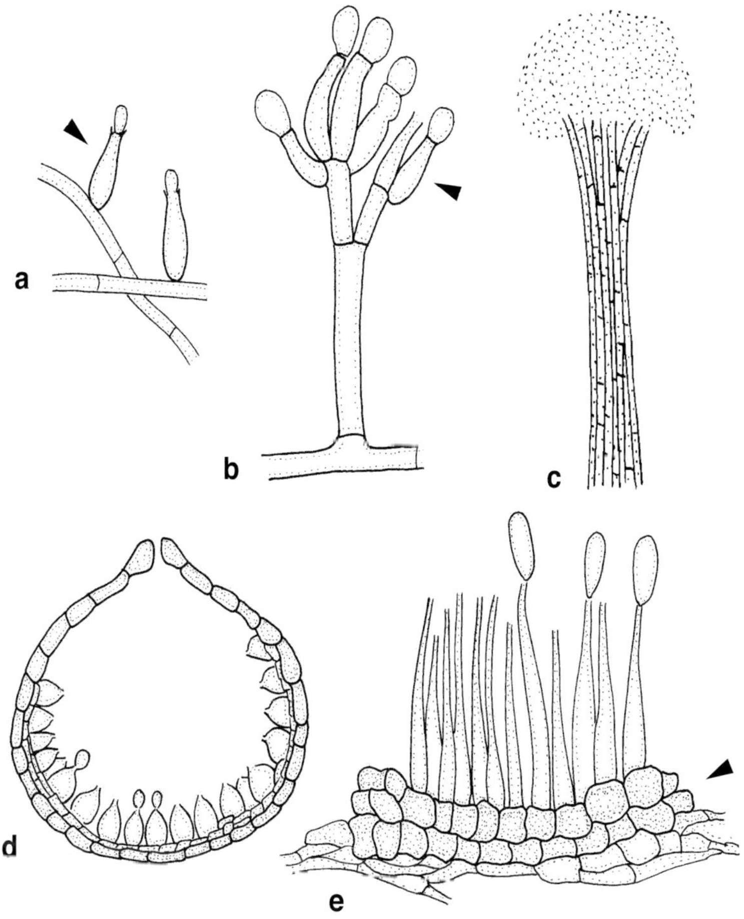

Fig. VII.3.1. Anamorphs of Ascomycota and Basidiomycota

a. Conidiogenous cells (arrow) formed directly on the mycelium. **b.** Conidiogenous cells (arrow) formed on conidiophores. **c.** Conidiophores bundled in a synnema (coremium). **d, e.** Fruit bodies: **d.** Pycnidium. **e.** Acervulus; the hyphae at the basis of the conidiogenous cell layer form a cellular tissue (arrow).

3.1 Conidiogenesis (according to Kiffer & Morelet 1999)

a. Thallic conidiogenesis (Fig. VII.3.2a–d)

Conidia are formed by disintegration of a hypha into several small units that differentiate into spores.

a.1 Holothallic conidiogenesis (Fig. VII.3.2a)

The distal part of the hypha is separated from the proximal one by a septum and the wall of the cell thus formed becomes thicker. This is the typical conidiogenesis observed in the **dermatophytes** (human and animal pathogenic fungi with the ability to infect hair, nail and skin).

a.2 Holoarthric conidiogenesis (Fig. VII.3.2b)

The whole hypha breaks down in several conidia. All layers of the cell wall are involved in the conidium formation.

a.3 Enteroarthric conidiogenesis (Fig. VII.3.2c)

The whole hyphae break down into several conidia, but only the inner layer of the cell wall is involved in the conidiogenesis. Some hyphal parts do not form conidia and often remnants of the outer layers of the wall can be observed on the free conidia.

a.4 Meristematic conidiogenesis (Fig. VII.3.2d)

This is also called "retrogressive conidiogenesis". The tip of a hypha develops into a conidium; when the first conidium is formed, the second develops immediately below it and thus the conidiogenous hypha becomes shorter and shorter.

b. Blastic conidiogenesis (Fig. VII.3.2e–j, VII.3.3)

Conidia are produced after formation of a new cell wall, often by budding from a conidiogenous cell. The locus on which the conidium is formed is sometimes called conidiogenous locus. Conidiogenous cells may develop directly on the mycelium or on conidiophores. After conidium formation, the conidiogenous cell may continue to grow apically.

b.1 Holoblastic conidiogenesis (Fig. VII.3.2e)

The conidia burst at the tip of the conidiogenous cell and all cell wall layers are involved in conidium development. Several different kinds of holoblastic conidiogenesis are known, depending on how the conidiogenous loci and the conidiogenous cell develop further.

b.1.1. Sympoduloconidia (Fig. VII.3.2f)

The conidiogenous cell grows sub-apically after each conidium formation; the conidiogenous loci are thus sympodial.

b.1.2. Botryoconidia (Botryoblastoconidia, Fig. VII.3.2g)

The conidia are formed simultaneously by budding on several conidiogenous loci on the same conidiogenous cell.

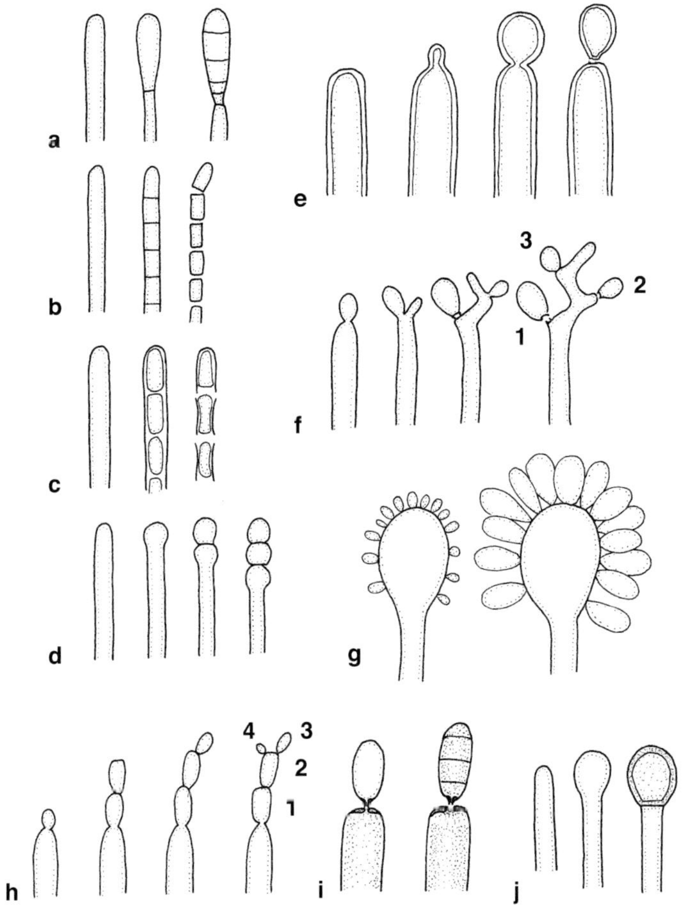

Fig. VII.3.2. Thallic and holoblastic conidiogenesis
a–d. Thallic conidiogenesis: **a.** Holothallic. **b.** Holoarthric. **c.** Enteroarthric. **d.** Meristematic arthroconidia. **e–j.** Holoblastic conidiogenesis: **e.** Holoblastic. **f.** Sympodial (1–3: sequence of the conidiogenesis). **g.** Botryoblastic. **h.** Acropetal chain (1–4: sequence of the conidiogenesis). **i.** Poroconidia. **j.** Aleurospores.

b.1.3. Acroblastic conidia (in acropetal chains, Fig. VII.3.2h)

The conidium itself becomes a conidiogenous cell, able to form one or more conidia, which in turn may give rise to other conidia. This produces so-called **acropetal chains** of conidia, in which the most recently formed conidium is at the top of the chain. Acropetal chains may be branched, while basipetal and thallic chains are not.

b.1.4. Poroconidia (Fig. VII.3.2i)

The conidiogenous locus is visible as a pore with a thick, melanised cell wall. This kind of conidium formation is mostly observed in fungi that produce dark-brown conidiophores, on which the pore is visible as a fine channel. Often poroconidia are characterised by the presence of a hilum, which is a scar-like structure that indicates the point of attachment of the conidium to the conidiogenous cell. Poroconidia may be formed singly or in acropetal chains.

b.1.5. Aleurospores (directly in the mycelium) / monoblastoconidia (in fruit bodies) (Fig. VII.3.2j)

Conidia are formed as a swelling of the tip of a differentiated conidiogenous cell or directly of an undifferentiated hypha. A septum is then separating the aleuroconidium from the conidiogenous cell. The conidium cell wall is thick and often multi-layered. Aleuroconidia and monoblastoconidia are released passively in the environment, i.e. they break loose from the conidiogenous cell only after death of the mycelium on which they are formed. Both can be considered resting organs, and correspondingly they may resist adverse environmental conditions. Ecologically, therefore, they are typical **memnospores**.

c. Enteroblastic conidiogenesis (Fig. VII.3.3)

Contrary to holoblastic conidiogenesis, enteroblastic conidiogenesis is characterised by the conidial cell wall being formed only by the inner layers of the conidiogenous cell walls. The conidiogenous cell can be compared to a bottle ("phiala" in Latin), from which conidia are released sequentially. The conidia so produced may stick together and collect at the opening of the conidiogenous cells as a droplet (Fig. VII.3.3b) or a chain (Fig. VII.3.3c), in which the youngest conidium is at the bottom (base): such chains are never branched and elongate at the base (**basipetal chains**).

c.1 Phialoconidia (Fig. VII.3.3a)

Conidiogenous cells that do not elongate are called phialides (Fig. VII.3.3a). They have usually a very fine, often tapering tip. The conidia thus formed ("phialoconidia") have often a narrow, rounded or tapering base.

c.2 *Anelloconidia (Fig. VII.3.3d)*

At the time of formation of each conidium, the conidiogenous cells may grow percurrently (i.e., growing continuously) and elongate, thus forming ring-like, sometimes conspicuous, more often, however, difficult to observe structures at the opening that gave rise to the name (**anellide**) for this kind of conidiogenous cell. Usually, anelloconidia are easily recognised by their broad, flattened base. They very often form basipetal chains.

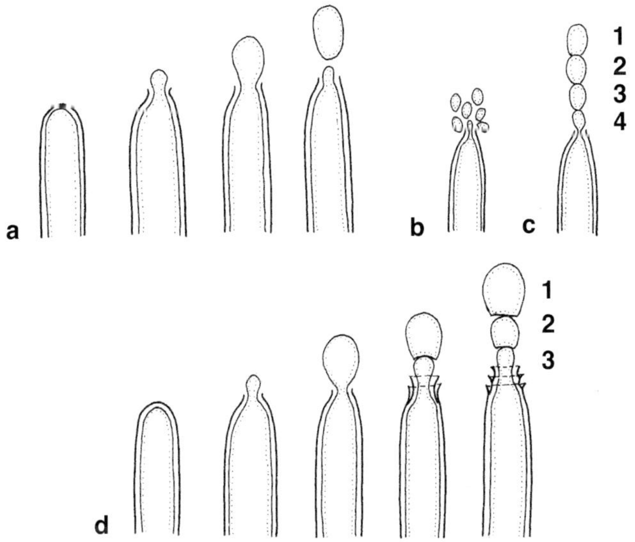

Fig. VII.3.3. Enteroblastic conidiogenesis

a–c. Enteroblastic conidiogenesis: **a.** Enteroblastic (phialidic). **b.** Formation of droplets. **c.** Basipetal chain of conidia (1–4: sequence of conidia formation). **d.** Anellidic (1–3: sequence of conidia formation).

Understanding conidiogenesis is crucial for a reliable identification of anamorphic fungi, yet it may be difficult for the beginner to memorise and distinguish the different types. Fig. VII.3.4 provides a schematic help to it.

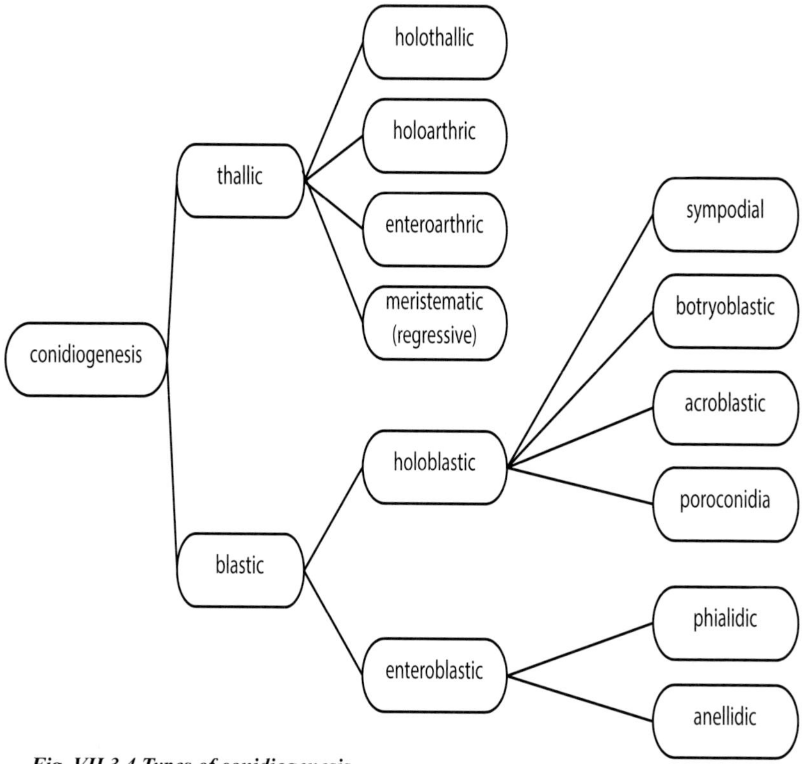

Fig. VII.3.4 Types of conidiogenesis

4. Spores: colour and shape (Fig. VII.4)

Spore shape and colour are important for the taxonomic distinction of fungal taxa.

Unpigmented, hyaline (colourless) spores are collectively called **hyalospores**; those with pigments, often brown, are named **phaeospores**. The spore colour should always be determined by observing spores preferably when they are in a group, because only slightly coloured spores may falsely appear as colourless if observed individually.

One-celled spores are sometimes called **amerosporous**, two-celled spores are **didymosporous** and those many-celled **phragmosporous**. Muriform (**dictyosporous**) spores are characterised by longitudinal and transverse septa. Dictyosporous spores are typical of phaeosporous taxa and are often observed in bitunicate genera.

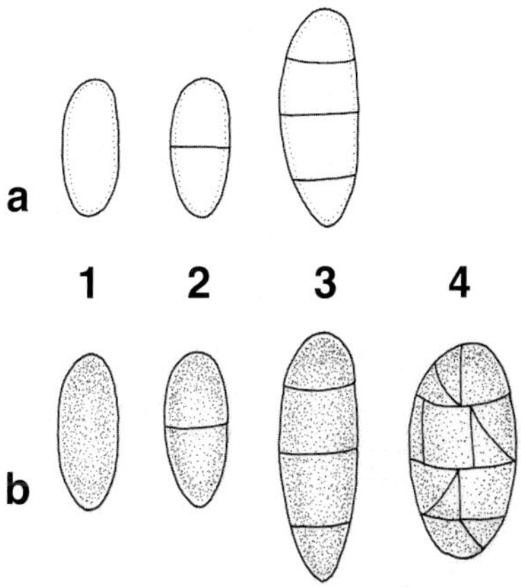

Fig. VII.4. Spore shape and colour

a. Hyalosporae (non-pigmented). **b.** Phaeosporae (pigmented). 1. Amerosporae (one-celled). 2. Didymosporae (two-celled). 3. Phragmosporae (many-celled). 4. Dictyosporae (muriform).

5. Conidial chains

Conidia may form chains of different types, discussed in detail in the keys. In Fig. VII.5 we provide a schematic definition of the different kinds of chains conidia may form.

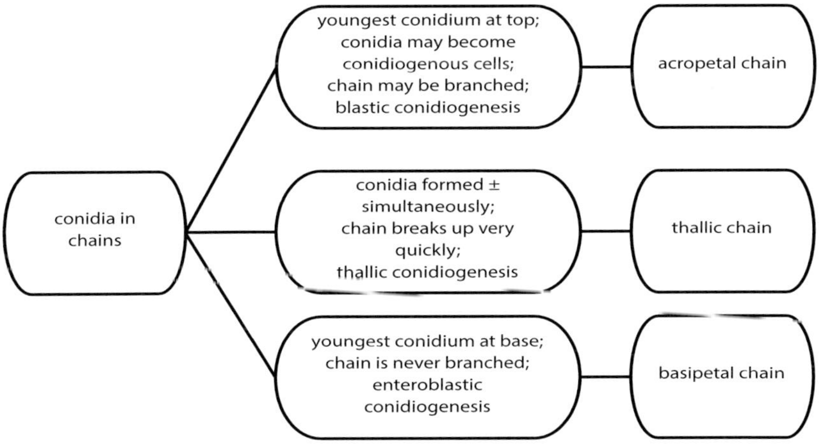

Fig. VII.5. Conidial chain types

VIII. Identification keys

1. Identification of moulds sporulating in pure culture

This is a simplified key for the identification of food-borne fungi and selected human and plant pathogenic moulds. It is intended to help beginners find their way in the complicated taxonomic world of microscopic fungi and should allow identification of moulds to the genus.

A necessary condition for a successful identification of moulds is the use of pure cultures and the rigorous recording of all cultural and microscopic characteristics. All observations should then be carefully documented with good drawings of critical features (in the digital era, this sounds somewhat obsolete – we prefer drawings to pictures, however, at least at the beginning, because one is forced to observe characters more accurately); this allows the mycologist to collect data that may be later used for comparison, at the same time building up mycological expertise.

The key includes only the most common genera of moulds and builds on the seminal works by von Arx (1981b) and Domsch et al. (2007). Genera not found in the following keys are probably rather uncommon and one has to refer to more specialised monographs. We also do not key out in detail some complex groups of genera such as the black yeasts and we refer to specialised literature instead. Identification to the species should then rely upon specialised monographs, which, if available, are reported for each genus in Section IX. De Hoog et al. (2000), St-Germain & Summerbell (2011), as well as Kiffer & Morelet (1997, 1999) provide additional identification keys to the genus and partly to selected species. The recently published, second edition of the classic "The genera of Hyphomycetes" (Seifert et al. 2011) can be considered the standard reference work for those mycologists dealing with hyphomycetous anamorphic fungi, but its use will most likely be reserved to advanced mycologists. The books by De Hoog et al. (2000) and St-Germain & Summerbell (2011) are recommended for the clinical mycologists and provide excellent illustrations of medically relevant moulds. The recently published "CBS Laboratory Manual Series 2: Food and indoor fungi" (Samson et al. 2011) is an invaluable tool for food and environmental mycologists.

The mycological taxonomic literature is unfortunately very rich in technical definitions. We provide a glossary at the end of this book that includes some of the most relevant mycological terms. A more comprehensive dictionary can be found in Ulloa & Hanlin (2000) and in the almost encyclopaedic "Dictionary of the Fungi (Kirk et al. 2011). The Internet is also a good place in which one can find common mycological terms, but definitions are sometimes wanting. Fungi of Australia (http://www.anbg.gov.au/glossary/webpubl/fungloss.htm#e) and Mycology Online (http://www.mycology.adelaide.edu.au/virtual/glossary/#e) provide quite careful and exact definitions of mycological terms. The official web site of the Mycoses Study Group, Doctor Fungus (http://www.doctorfungus.org/) is a very useful compilation of information on clinical mycology, with reasonably good keys to pathogenic species. Those interested in Zygomycota will probably find the best information (with excellent keys!) on this fascinating group of fungi in a dedicated site (http://www.zygomycetes.org/).

2. Identification of moulds: decision tree

This decision tree considers all the information to be collected before starting any identification.

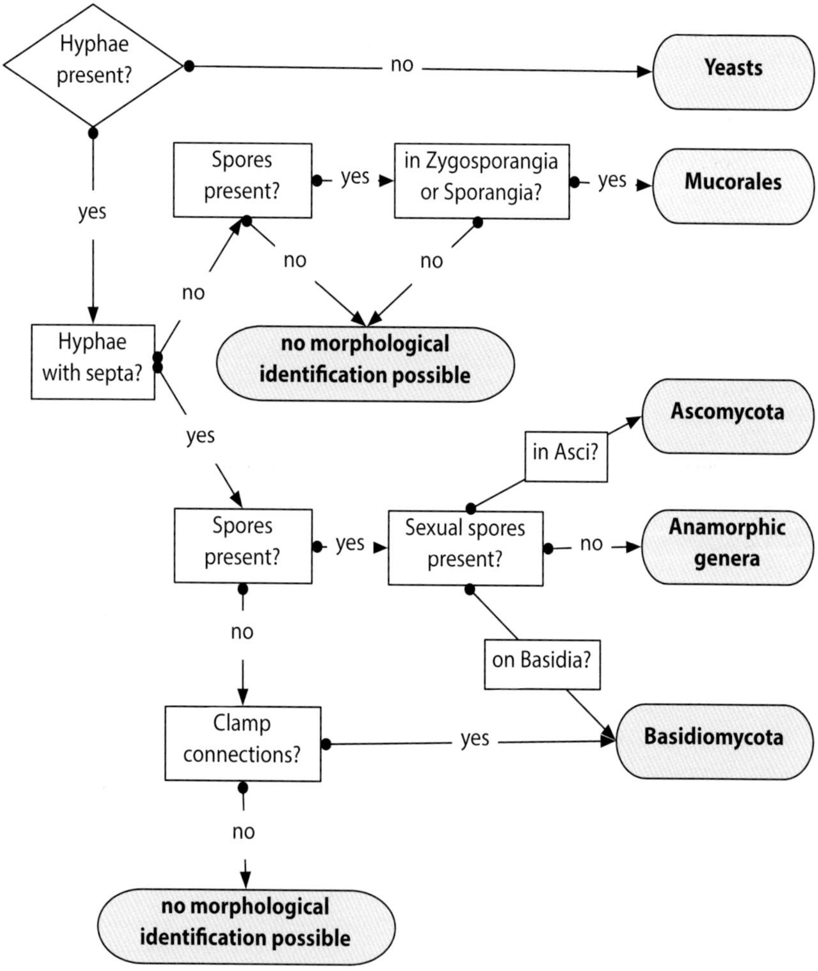

3. Main key

1. Zoospores (spores with flagella) present .... **Protista**
(specialised literature needed)

1.* Zoospores absent 2

2. Colonies formed by one-celled individuals, at most pseudomycelium present .... **Yeasts and some fungal Protista**
(specialised literature needed)

2.* True mycelium present 3

3. Spores absent .... **identification usually not possible**

3* Spores present.. 4

4. Hyphae with no septa [= coenocytic mycelium, with no transverse septa: some Zygomycota, however, may have some septate vegetative hyphae, particularly if the cultures are old. Zygomycetous fungi, on the other hand, have rather broad hyphae (up to 20 µm diam.), as opposed to Ascomycota, Basidiomycota and their anamorphs]. Teleomorph, if present, a zygosporangium containing one zygospore; anamorph a sporangium with endogenously produced sporangiospores .... **Zygomycota, Mucorales**

4. * Hyphae with septa 5

5. Spores produced sexually (by meiosis) on basidia or in asci 6

5.* Spores formed asexually (by mitosis) on conidiogenous cells .... **anamorphic Ascomycota and Basidiomycota**

6. Spores formed exogenously on basidia .... **Basidiomycota**
(specialised literature needed)

6.* Spores formed endogenously in asci .... **Ascomycota**

3.1 Zygomycota (Mucorales, Mortierellales, Zoopagales) (Fig. VIII.1–4)

1. Sporangia many-spored, spherical, ellipsoid, ovate or pyriform, formed on erect sporangiophores with no apical swelling 2

1.* Sporangia (merosporangia) containing only few sporangiospores (in some cases only one), cylindrical or spherical, formed on an apical swelling (vesicle) of the sporangiophore 11

2. All sporangia with a distinct columella, usually containing a very large number of sporangiospores (more than 100) 3

2.* Sporangia with an indistinct or without columella; if many-spored terminal sporangia with columella are seen, then sporangioles with no columella and containing less than 10 sporangiospores present 10

3. Zygosporangia more frequent than sporangia, with clearly asymmetrical, unequal zygophores with no appendages .... ***Zygorhynchus***

3.* Zygosporangia not more frequent than sporangia or absent; if present, zygophores are more or less equal 4

4. Sporangia with apophysis (funnel-shaped enlargement at transition from sporangiophore to sporangium).. 5

4.* Sporangia without apophysis 7

5 Sporangia globose, sporangiophores brown, arising opposite of rhizoids (root-like structures), sporangiospores irregular, often striate .... ***Rhizopus***

5.* Sporangia pear-shpead, sporangiophores hyaline, not arising opposite of rhizoids, sporangiospores regularly formed, smooth walled 6

6. Zygosporangium enveloped by appendages originating from the zygophores; mesophilic species, not growing above 37 °C .... ***Absidia***

6.* Zygosporangium not enveloped by appendages originating from the zygophores; thermophilic species, growing at or above 37 °C .... ***Lichtheimia***

7. Stolons (runners) with rhizoids present, sporangiophores arising opposite the rhizoids 8

7.* Stolons not formed, sporangiophores originating directly from the vegetative hyphae 9

8. Sporangiophores and sporangia hyaline to light-coloured, zygosporangia not known, mesophilic species .... ***Actinomucor***

8.* Sporangiophores and sporangia light brown, zygosporangia often present, thermophilic .... ***Rhizomucor***

9. Sporangiophores very tall (mostly more than 8 cm) and thick, with a bluish or metallic tinge, unbranched .... ***Phycomyces***

9.* Sporangiophores usually shorter than 3 cm, with no bluish or metallic tinge, often branched .... ***Mucor***

10. Sporangioles with persistent wall, without columella, containing only few (usually 1–4) sporangiospores, formed at the apex of dichotomously branched sporangiophores that arise laterally from the main sporangiophore, which may end with a terminal, many-spored sporangium with columella .... ***Thamnidium***

10.* Sporangioles with deliquescent wall, without or with a very small columella, containing up to 40 spores, formed apically on branched or unbranched sporangiophores .... ***Mortierella***

11. Merosporangia spherical, spiny, containing only one sporangiospore, produced all around a vesicle (swelling) of the sporangiophore .... ***Cunninghamella***

11.* Merosporangia cylindrical, containing several sporangiospores that are in a more or less distinct row.. 12

12. Several cylindrical merosporangia formed over the surface of a vesicle on a usually branched sporangiophore .... ***Syncephalastrum***

12.* Only few cylindrical merosporangia on the surface of the fertile vesicle, sporangiophores dichotomously branched; usually parasites of Mucorales .... ***Piptocephalis***

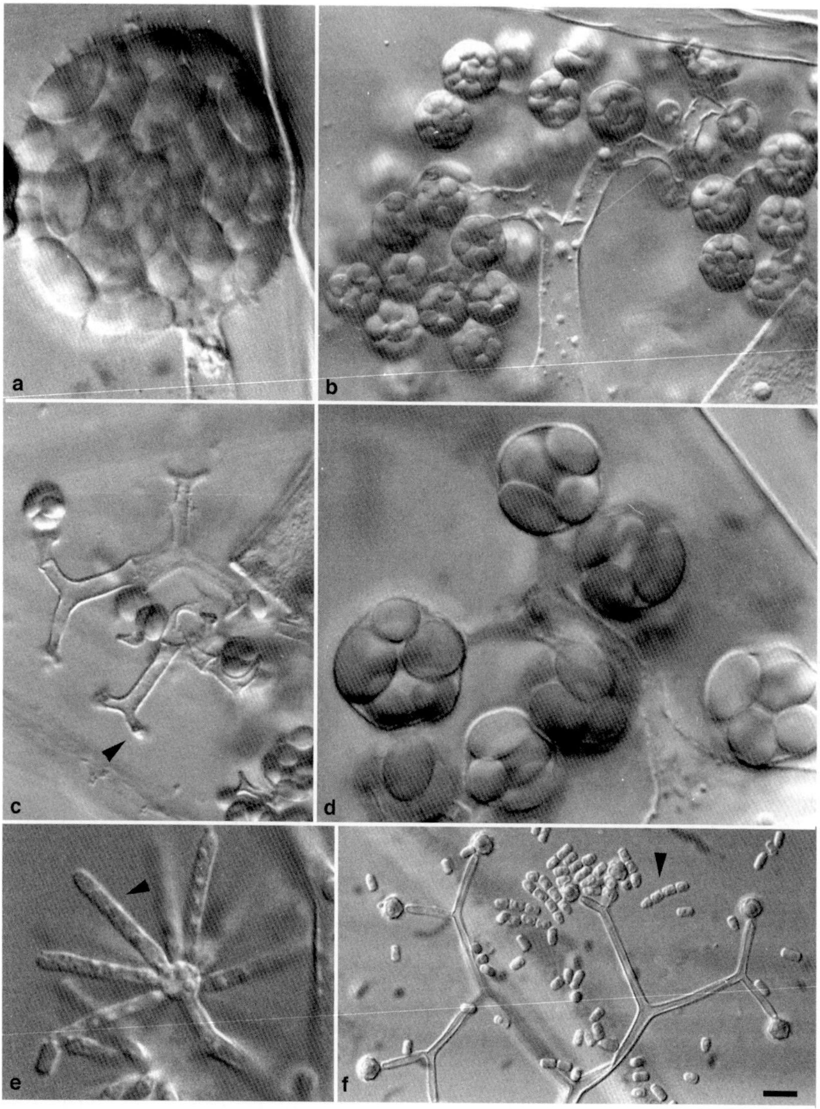

Fig. VIII.1. Zygomycota – 1

a–d. *Thamnidium elegans*: **a.** Sporangium with sporangiospores and intact sporangial membrane. **b.** Dichotomously branched sporangiole-bearing structures. **c.** Sporangiole-bearing structures from which sporangioles have detached (arrow). **d.** Sporangioles with spores. **e, f.** *Piptocephalis* sp.: **e.** Vesicle with merosporangia (arrow). **f.** Dichotomously branched sporangiophore; sporangiospores (arrow). Scale: b, c: 10 µm; f: 8 µm; a, d, e: 4 µm.

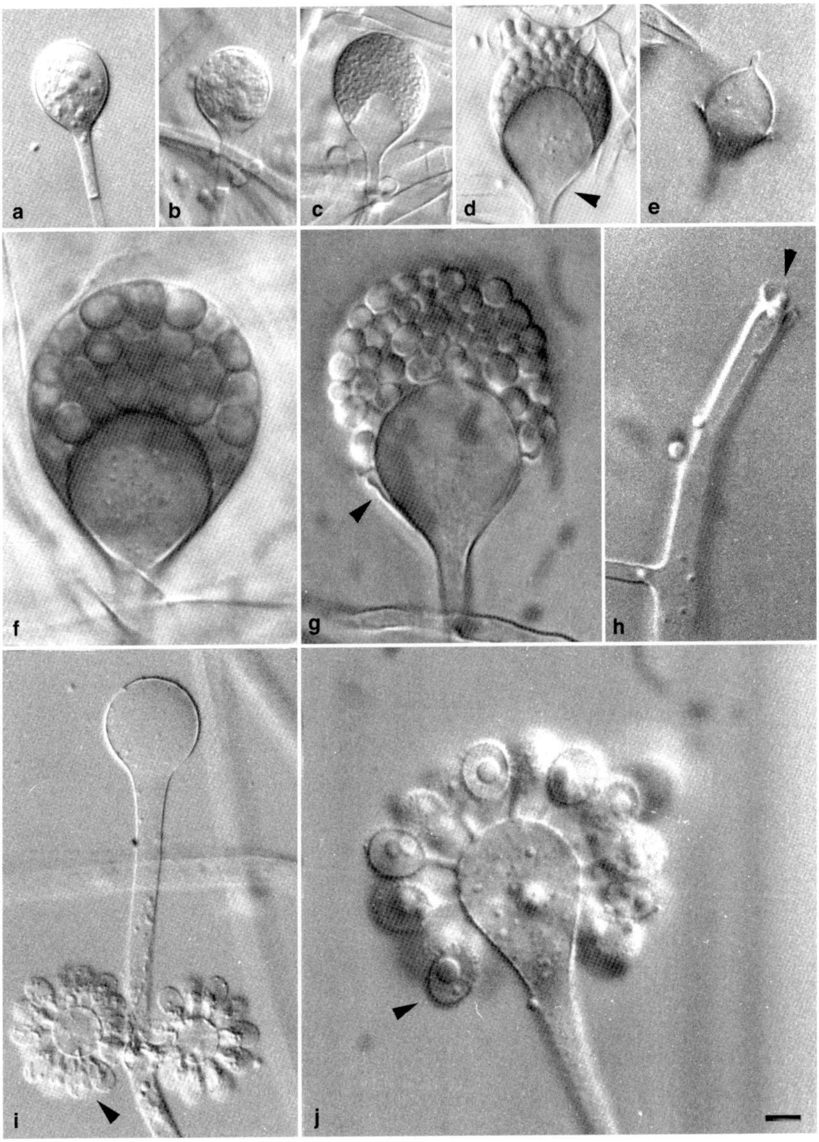

Fig. VIII.2. Zygomycota – 2

a–g. *Absidia cylindrospora*: **a–d.** Sporangiospore formation in the sporangium and spore discharge. Sterile central part of sporangium: columella with apophysis (arrow). **e.** Columella with apophysis. **f.** Sporangium. **g.** Sporangium with broken membrane (arrow). **h.** *Mortierella* sp.: reduced (almost absent) columella (arrow). **i, j.** *Cunninghamella elegans*: **i.** Sporangiophore with a large, terminal vesicle and smaller vesicles disposed in whorls around the terminal one (arrow). **j.** Vesicle with sporangioles containing only one spore (arrow). Scale: a–e, i: 10 µm; f–h, j: 4 µm.

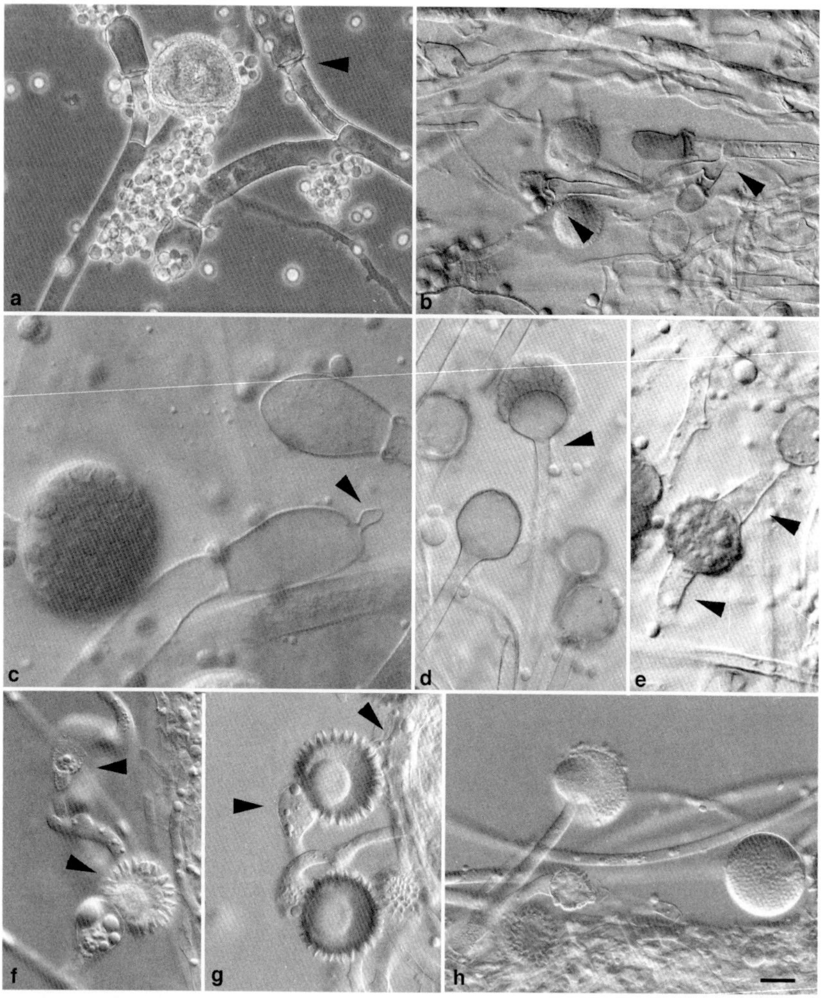

Fig. VIII.3. Zygomycota – 3

a, b. *Mucor fuscus*: **a.** Sporangium, columella without apophysis (arrow); **b.** Sympodial branching of the sporangiophore (arrow). **c.** *Mucor plumbeus*: Columella without apophysis, with a small peg typical of the species (arrow); on the left: ripe sporangium. **d, e.** *Rhizomucor miehei*: **d.** Sporangium and columella without apophysis (arrow). **e.** Symmetrical zygophores (arrows) with young zygosporangium. **f–h.** *Zygorhynchus moelleri*: **f.** Developing zygosporangia (arrows) with asymmetrical zygophores. **g.** Young zygosporangia with asymmetrical zygophores (arrows). **h.** Anamorphic *Zygorhynchus* (*Mucor*), aseptate hyphae. Scale: a, b, f–h: 20 µm; c–e: 10 µm.

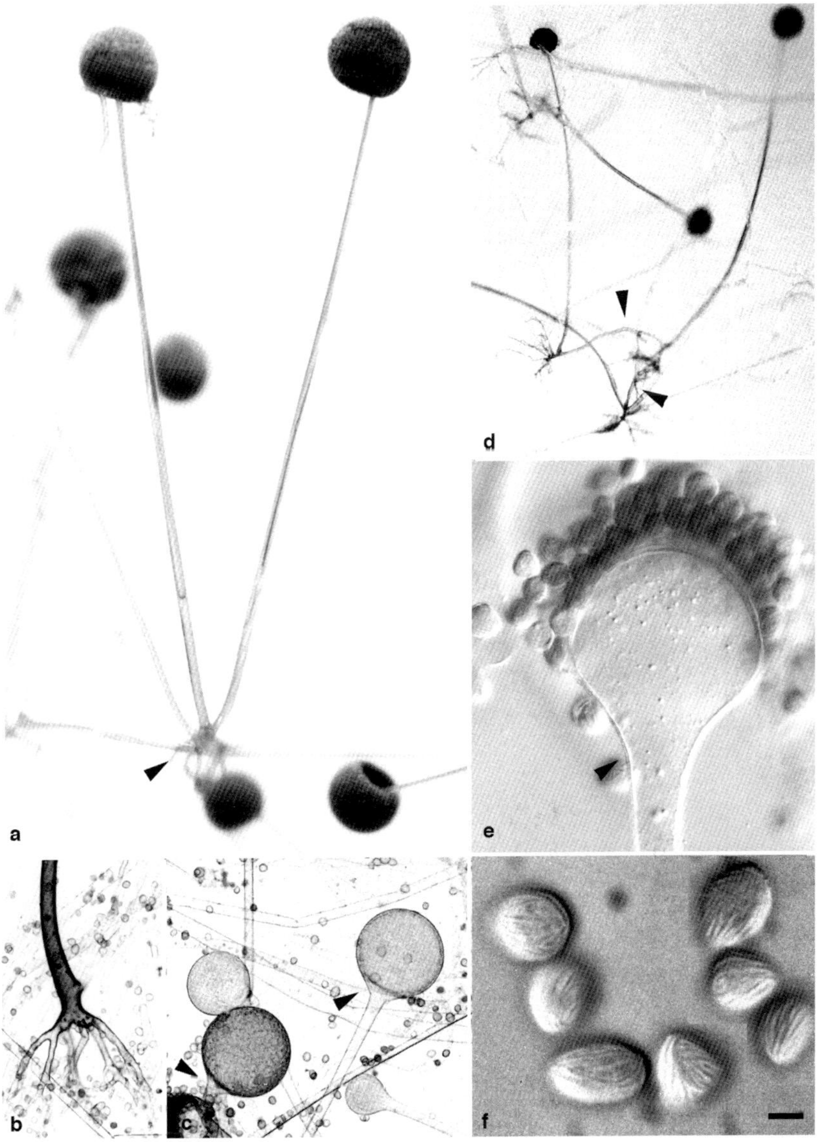

Fig. VIII.4. Zygomycota – 4

a–f. *Rhizopus stolonifer*: **a.** Sporangiophores with sporangia and rhizoids (arrow). **b.** Rhizoids. **c.** Sporangia with apophysis (arrows). **d.** Stolons (runners; arrows). **e.** Columella with apophysis (arrow), sporangiospores. **f.** Sporangiospores with ornamented walls. Scale: d: 200 μm; a: 100 μm; b, c: 40 μm; e: 10 μm; f: 4 μm.

3.2 Ascomycota (Fig. VIII.5–VIII.8)

1. Asci spherical or ellipsoid, with a thin wall that quickly dissolves when the ascospores ripen (prototunicate asci); ascomata are gymnothecia or cleistothecia 2

1.* Asci ellipsoid, club-shaped, bag-shaped, or cylindrical with often a pedunculate base, wall persistent well after the ascospores are ripe (unitunicate or bitunicate asci); ascomata are perithecia, apothecia or pseudothecia 19

Prototunicate asci [1]

2. Ascomata small (almost not discernible under a dissecting microscope), stalked, with few asci; cultures often reddish or greyish; anamorph *Basipetospora* .... ***Monascus***

2.* Ascomata large, not stalked, no *Basipetospora* anamorph 3

3. Ascospores usually hyaline or purplish-red, conidia produced enteroblastically (from phialides) in basipetal chains; anamorph *Aspergillus*, *Paecilomyces*, *Penicillium* or *Scopulariopsis* 4

3.* Ascospores usually yellow to brownish, conidia, if present, not in basipetal chains and produced through arthric, thallic or blastic conidiogenesis 14

Conidia in basipetal chains [3]

4. Ascomata not well developed, composed of loosely interwoven hyphae or at most of an inconspicuous, hyphal tissue; anamorph *Paecilomyces* .... ***Byssochlamys***

4.* Ascomata well developed, closed, with a conspicuous wall (peridium) composed of densely interwoven hyphae or of a cellular structure 5

5. Ascomata soft, single, peridium formed by a loose network of hyphae, asci catenulate (in short chains) .... ***Talaromyces***

5.* Ascomata hard, single or aggregated in a crust, peridium compact, cellular, asci not catenulate 6

6. Ascomata flattened, often aggregated in a crust-like, brown structure ("stroma"). Thermophilic, anamorph *Paecilomyces* .... ***Thermoascus***

6.* Ascomata more or less globose, usually not in stroma-like structures but sometimes in conglomerates; anamorph *Aspergillus*, *Penicillium* or *Scopulariopsis* 7

7. Anamorph *Aspergillus* 8

7.* Anamorph *Penicillium* or *Scopulariopsis* 13

8. Ascomata embedded in a stromatic tissue .... ***Dichlaena, Hemicarpenteles, Neopetromyces, Petromyces***

8.* Ascomata not embedded in a stromatic tissue 9

9. Ascomata surrounded by sterile, thick walled cells ("hüllecells") 10

9.* Ascomata naked, not surrounded by hüllecells 11

10. Ascomata surrounded by spherical or ellipsoid hüllecells, ascospores hyaline or yellowish, orange-reddish, purplish-red or blue .... ***Emericella***

10.* Ascomata surrounded by elongate or spiral hüllecells, ascospores hyaline or yellowish .... ***Fennellia***

11. Anamorph *Aspergillus* with globose heads; osmophilic species .. ***Eurotium***

11.* Anamorph *Aspergillus* with columnar heads; non-osmophilic species...... 12

12. Ascomata white .... ***Neosartorya***

12.* Ascomata yellow to brown .... ***Neocarpenteles***

13. Ascomata hard, sclerotised, without ostiolum, peridium pseudoparenchymatous, with thick-walled cells, ascospores rough-walled, symmetrical, often not produced; anamorph *Penicillium* .... ***Eupenicillium***

13.* Ascomata soft, carbonaceous, with an ostiolum, ascospores smooth-walled, often clearly asymmetrical (bean-shaped, triangular, etc.); anamorph *Scopulariopsis* .... ***Microascus***

Conidia not in basipetal chains [3*]

14. Ascomata a gymnothecium (peridium composed of a loose network of hyphae) 15

14.* Ascomata with a cellular peridium 16

15. Hyphae of peridium thin-walled, brownish, without appendages; ascospores ellipsoid or spindle-shaped .... ***Pseudogymnoascus***

15.* Hyphae of peridium thick-walled, colourless to brownish, stiff, with short and stout appendages; ascospores ellipsoid, lenticular or elongate .... ***Gymnoascus***

16. Ascomata glabrous 17

16.* Ascomata setose or hairy 18

17. Ascospores with one polar germ pore .... ***Thielavia***

17.* Ascospores with 2 polar germ pores .... ***Melanospora***

18. Hairs of ascoma with short, colourless appendages; ascospores with an equatorial furrow; conidia present .... ***Ascotricha***

18.* Hairs of ascoma without appendages but sometimes spiral or more or less dichotomously branched; ascospores with one, rarely two polar germ pores; conidia absent or very inconspicuous .... ***Chaetomium***

Unitunicate or bitunicate asci [1*]

19. Ascomata without ostiolum; asci with a double wall (bitunicate: a thin, inextensible outer wall covering a thick, elastic inner wall), club-shaped or bag-shaped; ascospores colourless to brown, with transverse and often longitudinal septa 20

19.* Ascomata with an ostiolum; asci with one wall (unitunicate: a single wall), usually cylindrical; ascospores without longitudinal septa, often one-celled 21

Bitunicate asci [19]

20. Ascospores with transverse and longitudinal septa, colourless to brownish.. .... ***Leptosphaerulina***

20.* Ascospores with only one transverse septum, colourless .... ***Didymella***

Unitunicate asci [19*]

21. Ascomata immersed in a light-coloured, hyphal stroma; ascospores colourless, yellowish, greenish, two-celled, disintegrating when ripe and thus asci apparently with 16 one-celled ascospores; anamorph *Trichoderma* .... ***Hypocrea***

21.* Ascomata free, not in a stroma 22

22. Ascomata pigmented, brown to black; ascospores dark, one celled, with a polar germ pore or a germ slit 23

22.* Ascomata light coloured (yellow, red, bluish); ascospores light coloured, many-celled, without germ pore or germ slit 26

23. Ascospores smooth 24

23.* Ascospores ornamented 25

24. Ascomata setose, ascospores with a germ slit .... ***Coniochaeta***

24.* Ascomata glabrous, ascospores with a polar germ pore .... ***Sordaria***

25. Ascospores with longitudinal stripes (striae) and germ pores ... ***Neurospora***

25.* Ascospores reticulate, with 1–2 germ pores .... ***Gelasinospora***

26. Ascospores one-celled .... ***Glomerella***

26.* Ascospores many-celled 27

27. Ascospores two-celled; anamorph *Gyrostroma*, *Tubercularia* or *Zythiostroma* .... ***Nectria*** complex

27.* Ascospores with more than two cells; anamorph *Fusarium* .... ***Gibberella***

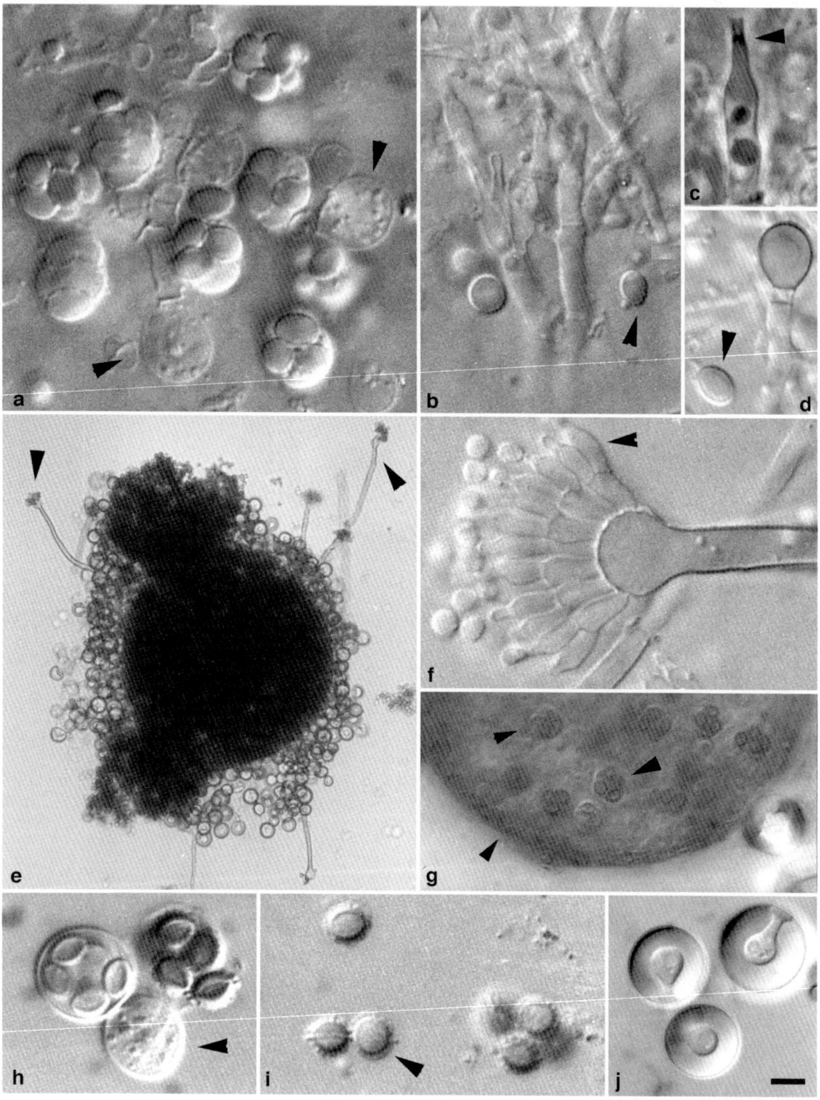

Fig. VIII.5. Ascomycota – 1

a–d. *Byssochlamys nivea*: **a.** Immature (arrow) and mature asci with ascospores. **b.** Anamorph (*Paecilomyces*): conidiophore, conidia (arrow). **c.** Conidiogenous cell (phialide) with collarette (arrow). **d.** Thick-walled ascospore and dark aleurospore. **e–j.** *Emericella nidulans*: **e.** Cleistothecium (in the central part) surrounded by hüllecells, with *Aspergillus* anamorphs (arrows). **f.** Anamorph (*Aspergillus nidulans*): conidiophore, conidiogenous cells (arrow), conidia. **g.** Peridium (arrow at bottom) of cleistothecium, asci (arrows at top) with ascospores. **h.** Immature ascus (arrow). Asci with immature and mature ascospores (clockwise). **i.** Ascospores with equatorial rings (arrow) and ornamented surface. **j.** Hüllecells. Scale: e: 40 μm; g: 10 μm; a–d, f, h–j: 4 μm.

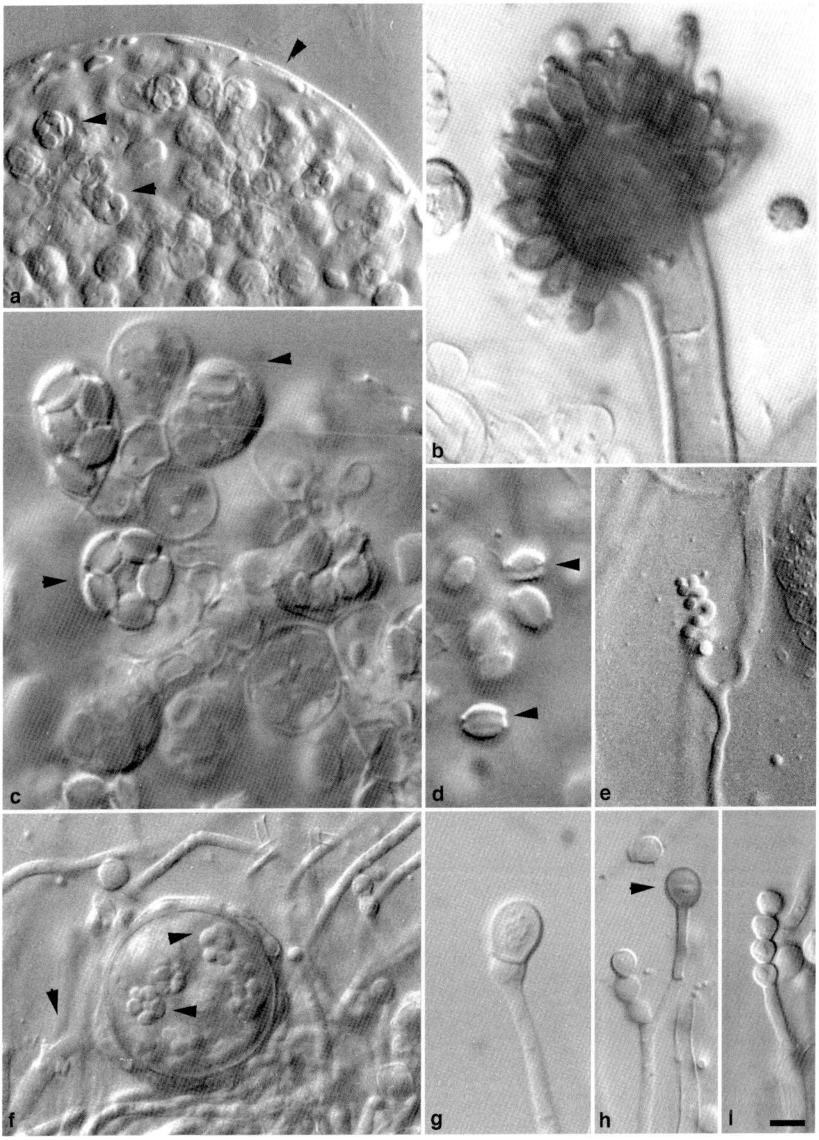

Fig. VIII.6. Ascomycota – 2

a–e. *Eurotium herbariorum*: **a.** Part of a cleistothecium; peridium (right arrow), asci (left arrows). **b.** Anamorph (*Aspergillus*). **c.** Immature (right arrow) and mature asci (left arrow) with thick-walled ascospores. **d.** Ascospores with equatorial furrow (arrows). **e.** Primordium (initials of a cleistothecium). **f–i.** *Monascus ruber*: **f.** stalked cleistothecium (left arrow), asci with ascospores (right arrows). **g–i.** Anamorph (*Basipetospora*: meristematic conidiogenesis), aleurospore (arrow). Scale: a, e, f, h, i: 1 µm; b–d, g: 4 µm.

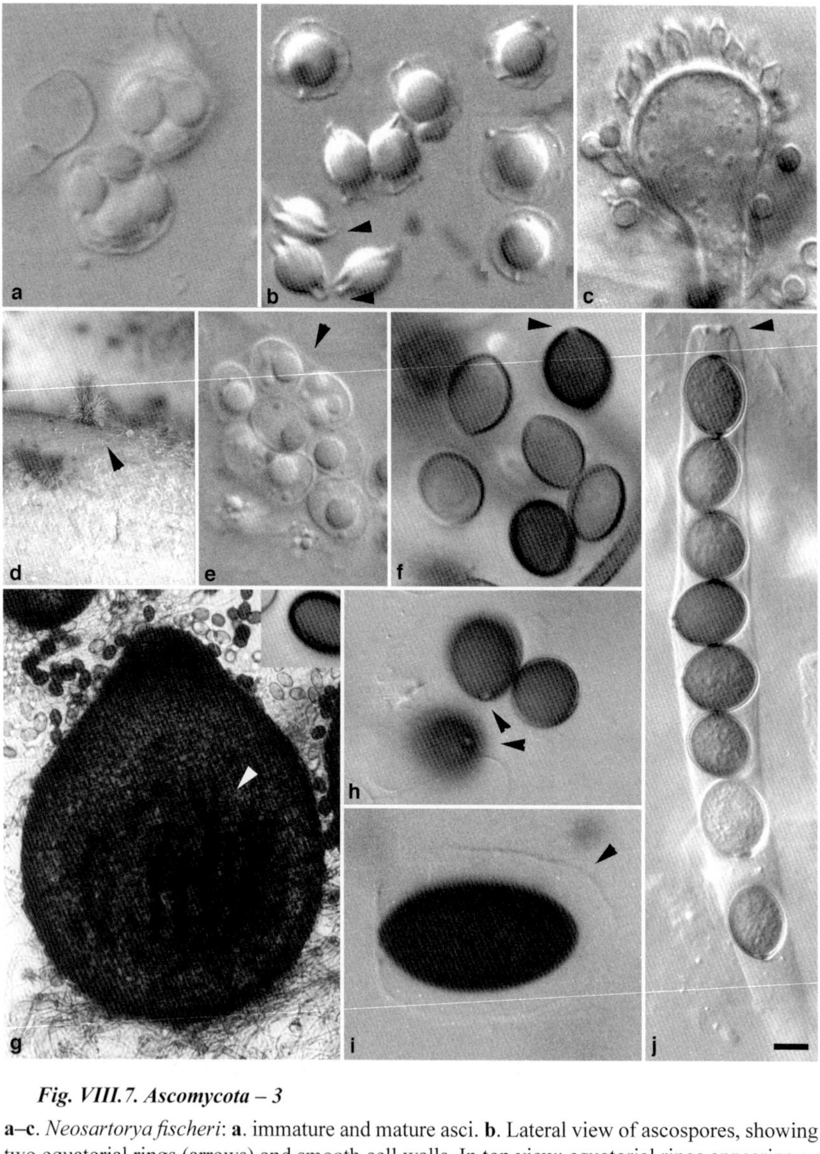

Fig. VIII.7. Ascomycota – 3

a–c. *Neosartorya fischeri*: **a**. immature and mature asci. **b**. Lateral view of ascospores, showing two equatorial rings (arrows) and smooth cell walls. In top view: equatorial rings appearing as a sheath. **c**. Anamorph (*Aspergillus*). **d–f**. *Chaetomium globosum*: **d**. Hairy ascoma (arrow). **e**. Ascus (arrow) with immature ascospores. **f**. Ascospores with germ pore (arrow). **g–j**. *Sordaria fimicola*: **g**. Perithecium with asci (arrow). **h**. Ascospores with germ pores (arrows). **i**. Ascospore with gelatinous sheath (arrow, water mount). **j**. Unitunicate ascus with ascus plug (arrow) and immature ascospores. Scale: d: 635 µm; g: 400 µm; h, j: 10 µm; a–c, e, f, i: 4 µm.

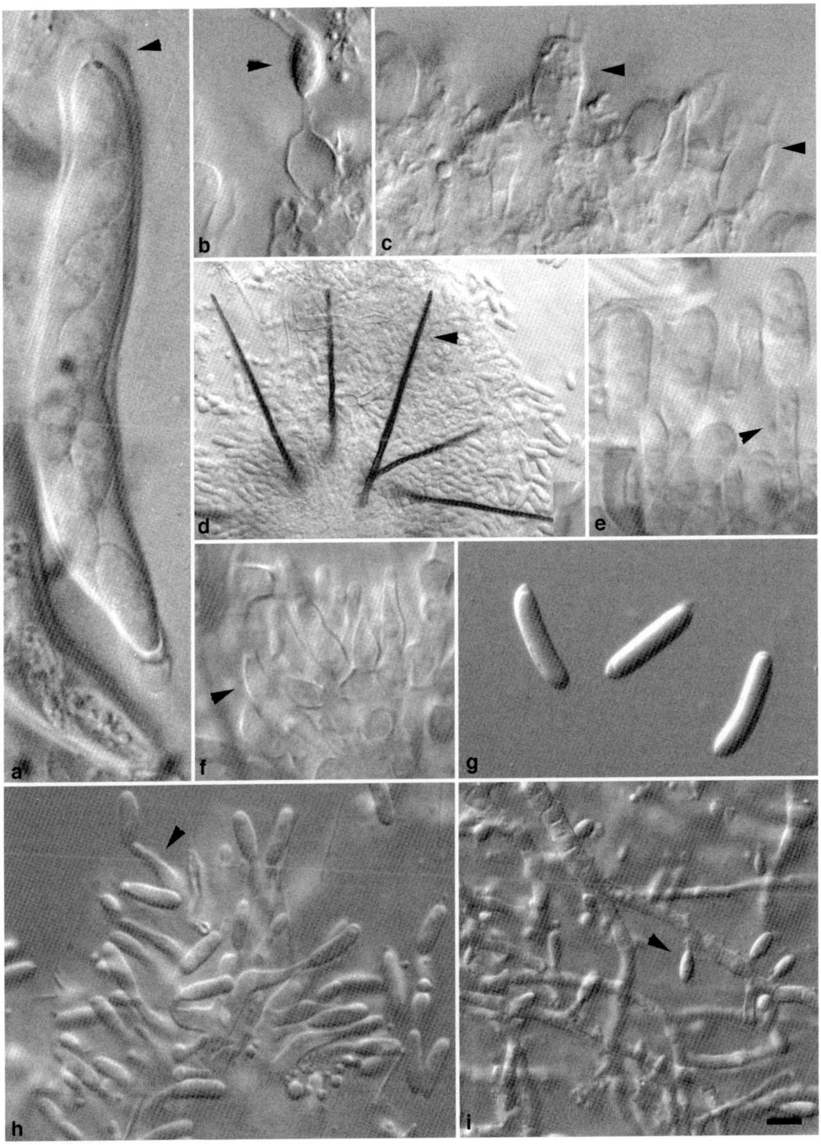

Fig. VIII.8. Ascomycota – 4, Basidiomycota, Coelomycetes

a. *Didymella bryoniae*: bitunicate ascus with two-celled ascospores. The two wall layers are particularly visible at the tip (arrow). **b, c.** Basidia (arrow, c) with basidiospore (arrow, b). **d, e.** *Colletotrichum*: **d.** *C. trichellum*: acervulus with setae (arrow). **e.** *C. gloeosporioides*: conidiogenous cells (arrow), conidia. **f, g.** *Cryptosporiopsis quercina*: **f.** Conidiogenous cells (arrow). **g.** Conidia. **h, i.** *Discula umbrinella*: **h.** Conidiogenous cells (arrow), conidia. **i.** Early developmental stage in culture, rudimentary conidiogenous cells. Scale: d, g, i: 10 µm; a–c, e, f, h: 4 µm.

3.3 Anamorphic Ascomycota and Basidiomycota

Many fungi, and correspondingly many moulds, produce only the anamorph in culture. In the mycological literature they should be reported more correctly by their teleomorph name, if known, or, as set out in the most recent nomenclatural rules, with the valid (accepted) name. In practice, however, the anamorph name is still and most likely will be widely used for quite some time. Section IX on page 89 reports the known anamorph-teleomorph connections for all anamorphic genera discussed here, along with the most relevant bibliographic references.

A. Conidiogenous cells produced on or in fruiting bodies (conidiomata) that can be flat, disc-shaped (acervuli) or closed, with a well-defined wall (pycnidia) .... **"Coelomycetes"**

B. Conidiogenous cells produced more or less all over the thallus, originating either directly from vegetative hyphae or born on more or less erect conidiophores that are distinct from the vegetative hyphae and sometimes may be bundled, forming so-called synnemata or sporodochia ... **"Hyphomycetes"**

3.3.1 Coelomycetes (Fig. VIII.8, 9)

1. Conidiogenous cells in pycnidia 2

1.* Conidiogenous cells in acervuli 10

2. Conidia brown, two-celled, more rarely one-celled, usually quite large (>20 µm) .... ***Diplodia***

see also *Botryodiplodia* and *Lasiodiplodia*

2.* Conidia hyaline, one celled or multiseptate 3

3. Conidia ellipsoid or flattened on one side, predominantly one-celled, usually small (<10 µm), without appendages or mucilaginous sheath 4

3.* Conidia spherical, vermiform, filiform or cylindrical, one celled or multiseptate, usually larger than 10 µm, sometimes with appendages and/or mucilaginous sheaths 5

4. Conidiogenous cells ampulliform, inconspicuous, conidia hyaline (if pale brown, cf. *Microsphaeropsis*), chlamydospores often present, cultures often dark-coloured .... ***Phoma***

4.* Conidiogenous cells filiform, elongate, conidia hyaline, often flattened on one side, chlamydospores absent, cultures greyish, light-coloured .... ***Hainesia***

5. Conidia one-celled 6

5.* Conidia two-celled or multiseptate 9

6. Conidia vermiform, borne on elongated conidiogenous cells, conidiomata often very open, reminiscent of an acervulus .... ***Phlyctema***

6.* Conidia filiform, fusiform or cylindrical, with or without appendages 7

7. Conidia ("beta conidia") filiform, curved; irregularly fusiform (alpha conidia) may also be present (in some cultures beta conidia are scant or absent) .... ***Phomopsis***

7.* Conidia cylindrical or spherical, with a mucilaginous appendage 8

8. Conidia cylindrical, with a fan-shaped mucilaginous appendage .... ***Ceuthospora***

8.* Conidia ellipsoid or spherical, with a filiform appendage and/or a mucilaginous sheath .... ***Phyllosticta***

9. Conidia one-septate, of constant length .... ***Ascochyta***

9.* Conidia multiseptate, of variable length .... ***Stagonospora***

10. Conidia hyaline, one-celled 11

10.* Conidia brown, one-celled or multiseptate 13

11. Conidiomata rudimentary, conidia ellipsoid to cylindrical, shorter than 12 µm .... ***Discula***

11.* Conidiomata distinct, conidia usually longer than 12 µm 12

12. Conidiomata flattened, often with bristle-like, brownish, soft structures (setae); conidia cylindrical, lunate, sickle-shaped or fusiform, often with a granular content, usually longer than 12 µm; appressoria often formed after conidia have been incubated a few hours in water .... ***Colletotrichum***

12.* Conidiomata pulvinate, large, without setae, with a basal, thick, dark layer of stromatic cells; conidia elongate-cylindrical, if fusiform then rather small but usually longer than 12 µm, in some species up to 30 µm; appressoria absent .... ***Cryptosporiopsis***

13. Conidia one-celled .... ***Melanconium***

13.* Conidia septate 14

14. Conidia muriform, warty, without appendages .... ***Epicoccum***

14.* Conidia with only transverse septa, with branched, hyaline apical appendages; terminal cells lighter pigmented than central cells ...... ***Pestalotiopsis*** group

3.3.2 Hyphomycetes (Fig. VIII.10–18)

1. Synnemata (coremia) formed by bundled conidiophores present 2

1.* Synnemata absent 4

2. Conidia hyaline, forming mucilaginous, white aggregates .... ***Graphium***

2.* Conidia pigmented (grey or green), usually catenulate 3

3. Synnemata grey or brown; conidiogenous cells originating from the upper part of the synnema, appressed; cultures and conidia brown or grey to black, relatively large, >5 µm in diam., with a truncate base, in basipetal chains (youngest conidium at the base of the chain: enteroblastic conidiogenesis) on anellides (conidiogenous cell with broad neck, elongating apically)............ .... ***Cephalotrichum*** (***Doratomyces***)

3.* Synnemata greenish; conidiogenous cells brush-like, erumpent from all over the synnematous structures; cultures and conidia green, usually <5 µm in diam., ellipsoid, without truncate base, in basipetal chains on phialides (conidiogenous cell with tapering neck, not elongating) .... ***Penicillium*** pr. p. (e.g. *P. vulpinum* group)

4. Conidia catenulate, conidiogenesis blastic or thallic.................................. 5

4* Conidia not catenulate (single or in droplets).. 20

Conidia catenulate [4]

5. Conidia in basipetal chains [youngest conidium at the base of the chain; enteroblastic or retrogressive (meristematic) conidiogenesis], chain unbranched. 6

5.* Conidia not in basipetal chains, chains may be branched 14

Basipetal or retrogressive chains [5]

6. Conidia borne in basipetal succession (enteroblastic conidiogenesis) from phialides (conidiogenous cell with fixed length, not elongating) or anellides (conidiogenous cells elongating).. 7

6.* Conidiogenesis retrogressive (= meristematic conidiogenesis) from conidiogenous hyphae (conidiogenous cell becoming shorter after each conidium) 12

7. Phialides with a long (≥10 µm), cylindrical tube-like neck; conidia borne endogenously within the cylindrical tube; catenulate or septate, dark pigmented chlamydospores present .... ***Thielaviopsis***

7.* Phialides with a short neck or no neck at all; conidia not born in a cylindrical tube but at the tip of the conidiogenous cell; chlamydospores (aleuroconidia), if present, one-celled, hyaline or light coloured 8

8. Phialides formed directly on apical vesicles (swelling ≥15 µm diam.; if <6 µm, cf. *Penicillium*) of unbranched conidiophores (monoseriate) or phialides borne on a row of sterile cells (metulae) disposed on the vesicles (biseriate) .... ***Aspergillus***

8.* Phialides forming brush-like verticillia on branches with narrow subtending angles, if brush-like verticillia absent, conidiophores with no or very small apical swelling (<6 µm) and lacking a row of sterile cells 9

9. Conidia with a truncate base, light-coloured to brownish, brown or black; conidiogenous cells are anellides, cylindrical or swollen at the base .... ***Scopulariopsis***

if conidia not in chains but in slimy droplets, see *Scedosporium*

9.* Conidia without truncate base, ellipsoid, usually greenish, yellowish or olive in mass, rarely brown, conidiogenous cells are more or less flask-shaped phialides 10

10. Conidiophore stipe brown .... ***Thysanophora***

10.* Conidiophore stipe light-coloured 11

11. Conidiophore branching brush-like only at the top (conidiophores may be branched throughout), quite regular; conidiogenous cells cylindrical or elongate, never abruptly tapering, arising directly from the conidiophores (mono-

verticillate) or forming verticillia on sterile branches (metulae), conidia green in mass .... ***Penicillium***

11* Conidiophore branching brush-like not only at the top, but also below, giving the conidiophore a rather irregular shape; conidiogenous cells usually awl-shaped (subulate, abruptly tapering becoming slender and tapering to a point); conidia yellowish, olive or brownish in mass .... ***Paecilomyces***

12. Conidia two-celled, both cells more or less equally large, basal cell beaked; conidia borne obliquely on long, unbranched, filamentous conidiogenous cells; culture pink .... ***Trichothecium***

12.* Conidia one-celled 13

13. Culture yellow-orange .... ***Basipetospora***

13.* Culture white, slow growing .... ***Fraseriella***

Thallic or acropetal chains [5*]

14. Conidia usually cylindrical, one-celled, formed by breaking up of the hyphae (thallic conidiogenesis) 15

14.* Conidia of various shapes, one-celled or septate, produced by budding of the conidiogenous cell (holoblastic conidiogenesis), forming acropetal (youngest conidium at the apex of the chain), often branched chains 17

15. Chains always composed of 4 light brown conidia that become spherical when ripe (ripening of conidia in basipetal succession) .... ***Wallemia***

15.* Chains elongating indefinitely, conidia always hyaline 16

16. Colonies white, appressed, dry or slimy, moderately rapidly growing (a 90 mm Petri dish covered within 3–5 days) .... ***Geotrichum***

16.* Colonies orange, fluffy, dry, extremely rapidly growing (a 90 mm Petri dish covered within 24 hours!) .... ***Chrysonilia***

17. Conidia one-celled, hyaline, spherical or barrel-shaped, chains with or without disjunctor cells (empty cells between two conidia); microconidia often present .... anamorphic ***Monilinia***

17.* Conidia one-celled or septate, brown, of various shapes 18

18. Conidia clavate, often rostrate, with longitudinal, transverse and oblique septa (muriform) .... ***Alternaria***

18.* Conidia not clavate, never rostrate, with only transverse septa or one-celled 19

19. Conidiophores well-developed, brown; conidia one-celled or with one- to four transverse septae, ellipsoid to fusiform, smooth or lightly warty, light brown, olive green in mass, not constricted at the septa; conidial chains mostly conspicuously branched but soon disintegrating .... ***Cladosporium***

19.* Conidiophores not formed and conidiogenous cells often not well defined; conidia with several transverse septae, cylindrical, warty, dark brown, constricted at the septae .... ***Torula***

Conidia not in chains [4*]

20. Conidiogenous cells are phialides; conidia forming slimy droplets around the opening of the conidiogenous cell 21

20.* Conidiogenous cells are no phialides; conidia produced holothallically, holoblastically or borne from anellides 34

Phialidic conidiogenesis [20]

21. Conidia brown or black, warty, one-celled; conidiophores brownish, evenly distributed over the whole thallus, bearing a terminal cluster of brown phialides that are widest at the apex .... ***Stachybotrys***

21.* Conidia hyaline or light-coloured, smooth or warty, one- or many-celled, conidiophores and phialides not as above 22

22. Conidia one-celled 23

22.* Conidia many-celled 30

23. Phialides arising directly from the vegetative hyphae, conidiophores absent, sometimes in small whorls on lateral branches of the main hyphae 24

23.* Phialides borne on well-defined conidiophores 27

24. Phialides often with a more or less conspicuous collarette, flask-shaped, brownish; culture and mainly also mycelium brown to black .... ***Phialophora*** complex

includes *Exophiala, Wangiella, Cladophialophora*

24.* Phialides with no or inconspicuous collarette, flask shaped, cylindrical or slender and tapering, hyaline; cultures and mycelium light-coloured (white, pink, yellow) 25

25. Phialides flask-shaped, with a swollen base .... ***Tolypocladium***

25.* Phialides not as above 26

26. Phialides cylindrical, slender and tapering towards the top, with reduced or no collarette, mostly hyaline; culture white, sometimes yellowish, reddish or greenish .... ***Acremonium***

cf. also *Fusarium* forms producing only microconidia

26.* Phialides usually sterigmata-like, reduced to a peg forming a single conidium (adelophialides), hyaline; culture white .... ***Aphanocladium***

27. Two types of conidiophore branching (verticillate and brush-like, or both on the same branch) present in the same culture .... ***Clonostachys***

27.* Two types of conidiophore branching not present 28

28. Conidiophores obligately verticillate, branches and upper parts of the conidiophore bearing verticillate whorls of phialides .... ***Verticillium***

28.* Conidiophores with no verticillate branching and without verticillate whorls of phialides 29

29. Branches with phialides in irregular whorls, not forming cushions; cultures white, yellowish, greenish or reddish (differs from *Penicillium* mainly by the lack of conidial chains); phialides, elongated, cylindrical .... ***Gliocladium***

29.* Branches with phialides more or less at right angles from the main axis, often in dense cushions; cultures white or green; phialides short, swollen, sometimes curved .... ***Trichoderma***

30. Conidia predominantly two-celled or at most three-celled 31

30.* Conidia two-celled to many-celled 32

31. Conidiophores are long, unbranched filamentous conidiogenous cells forming obliquely two-celled conidia with both cells more or less equally large and beaked basal cell; culture dry, pink .... ***Trichothecium***

31.* Conidiophores inconspicuous, short, branched, often on hyphal strands with short, conidiogenous cells inflated at the base forming two- to three-celled conidia without beak, cultures slimy, white to orange .... anamorphic ***Monographella***

32. Conidia elongate, sickle-shaped, with two or more transverse septa, with a distinct foot cell; one-celled microconidia often present (sometimes only the microconidia are formed in culture!); colonies white or with greenish, red, yellow or purplish tinge .... ***Fusarium***

32.* Conidia more or less cylindrical, two- to many-celled, without any distinct foot cell, with a rounded base; microconidia present or absent; colonies white to red-brown 33

33. Conidiophores long, erect, apically brush-like branched and forming a long, prominent stipe extension that terminates in a club-shaped vesicle; macroconidia cylindrical, septate; microconidia, if present, cylindrical .... ***Cylindrocladium***

33.* Conidiophores short, irregularily branched near the base, without stipe extension and terminal vesicle; macroconidia curved, septate; microconidia, if present, ellipsoid .... ***Cylindrocarpon***

Thallic, holoblastic or anellidic conidiogenesis [20*]

34. Conidia hyaline or light brownish 35

34.* Conidia dark brown 44

Conidia hyaline or light brownish [34]

35. Cultures slimy, yeast-like and light-coloured when young, later becoming dark; conidia one-celled, often budding; chlamydospores abundant, brown to black; a complex of different genera that are morphologically almost undistinguishable .... ***Aureobasidium***

"black yeasts"; see also *Exophiala, Hormonema, Ramichloridium, Rhinocladiella*

35.* Cultures farinose or with an aerial mycelium; conidia one-celled or septate, budding conidia not present; chlamydospores not abundant, if present, light coloured 36

36. Conidia with two unequal cells; upper, larger cell with a spinose surface; conidiophores inconspicuous .... ***Mycogone***

36.* Conidia not as above, one-celled to many-celled; conidiophores inconspicuous or well-developed 37

37. Conidiogenesis holoblastic, conidia one-celled 38

37.* Conidiogenesis thallic (holothallic) or anellidic 39

38. Conidiogenous cells are apical swellings of the conidiophore branches, forming many conidia simultaneously; conidia ellipsoidal, <15 µm wide .... ***Botrytis***

38.* Conidiogenous cells are elongated on inconspicuous conidiophores forming conidia singly; conidia spherical, very large (usually >30 µm) ***Sepedonium***

39. Conidia one-celled 40

39.* Conidia many-celled, sometimes one-celled microconidia present; often isolated from animal or human tissue (esp. skin); cultures often with an unpleasant and intense smell 42

40. Conidiogenesis anellidic – rings on the conidiogenous cells, however, inconspicuous; conidiogenous cells tapering; conidia hyaline or light brown, with a slightly flattened base, aggregated in small, slimy droplets around the opening of the anellide .... ***Scedosporium***

if conidia in chains, cf. *Scopulariopsis*

40.* Conidiogenesis holothallic 41

41. Conidia flattened at the base, often thick-walled and warty, usually shorter than 6 µm, borne on sometimes inconspicuous conidiophores and occasionally in short chains .... ***Chrysosporium***

41.* Conidia ovate, pyriform, clavate, sometimes match-stick-like, thin-walled, smooth, usually longer than 6 µm, borne directly on the mycelium; dermatophyte, human or animal pathogen .... ***Trichophyton***

42. Conidia (macroconidia) many-celled, with a rough cell wall surface, often spindle- or cigar-shaped, sometimes pyriform; one-celled microconidia sometimes present .... ***Microsporum***

42.* Conidia (macroconidia) many-celled, with a smooth cell wall surface; one-celled microconidia present or absent 43

43. One-celled microconidia absent; macroconidia clavate; culture green-brownish or yellow-orange .... ***Epidermophyton***

43.* One-celled microconidia present, predominant; macroconidia cylindrical to cigar-shaped .... ***Trichophyton***

Conidia dark [34*]

44. Conidia one-celled, born singly on rather undifferentiated conidiophores; small, spherical, one celled phialoconidia sometimes present 45

44.* Conidia many-celled, born on differentiated or undifferentiated conidiophores 47

45. Thermophilic (growing above 40 °C) .... ***Thermomyces***

45.* Mesophilic (growing below 37 °C) 46

46. Conidia with no germ slit, spherical, born on small cylindrical cells .... ***Humicola***

46.* Conidia with a germ slit, ellipsoid, borne directly on hyphae (if penicillate conidiogenous cells present cf. *Wardomyces*; if conidiogenous cells swollen cf. *Nigrospora*) .... ***Mammaria***

47. Conidia with transverse and longitudinal septa 48

47.* Conidia with transverse septa only, septa sometimes inconspicuous 50

48. Conidia attached to the conidiogenous cell with the whole, flattened base, when detached, base sometimes with remnants of the conidiogenous cell ... . ***Pithomyces***

48.* Conidia attached to the conidiogenous cell only by a reduced part, at the site of attachment often an annular scar visible 49

49. Conidiophores growing percurrently; conidia attached to the conidiogenous cell only by an exiguous pore .... ***Stemphylium***

49.* Conidiophores growing sympodially; conidia without pore; the conidium itself may become a conidiogenous cell or a conidiophore .... ***Ulocladium***

50. Conidia two-celled; conidiogenous cells cylindrical, with a swollen base (not always visible in culture), with conspicuous anellides .... ***Spilocea*** (anamorph of *Venturia*)

50.* Conidia many-celled: conidiogenous cells cylindrical, without swollen base, with pores or scars 51

51. Conidia elongate fusiform, often sickle-shaped, terminal cells sometimes lighter-coloured than median cells .... ***Curvularia***

51.* Conidia straight or minimally curved 52

52. Conidium base with a conspicuous scar (hilum) .... ***Exserohilum***

52.* Conidium base without hilum 53

53. Conidia mostly cylindrical .... ***Drechslera***

53.* Conidia mostly spindle-shaped .... ***Bipolaris***

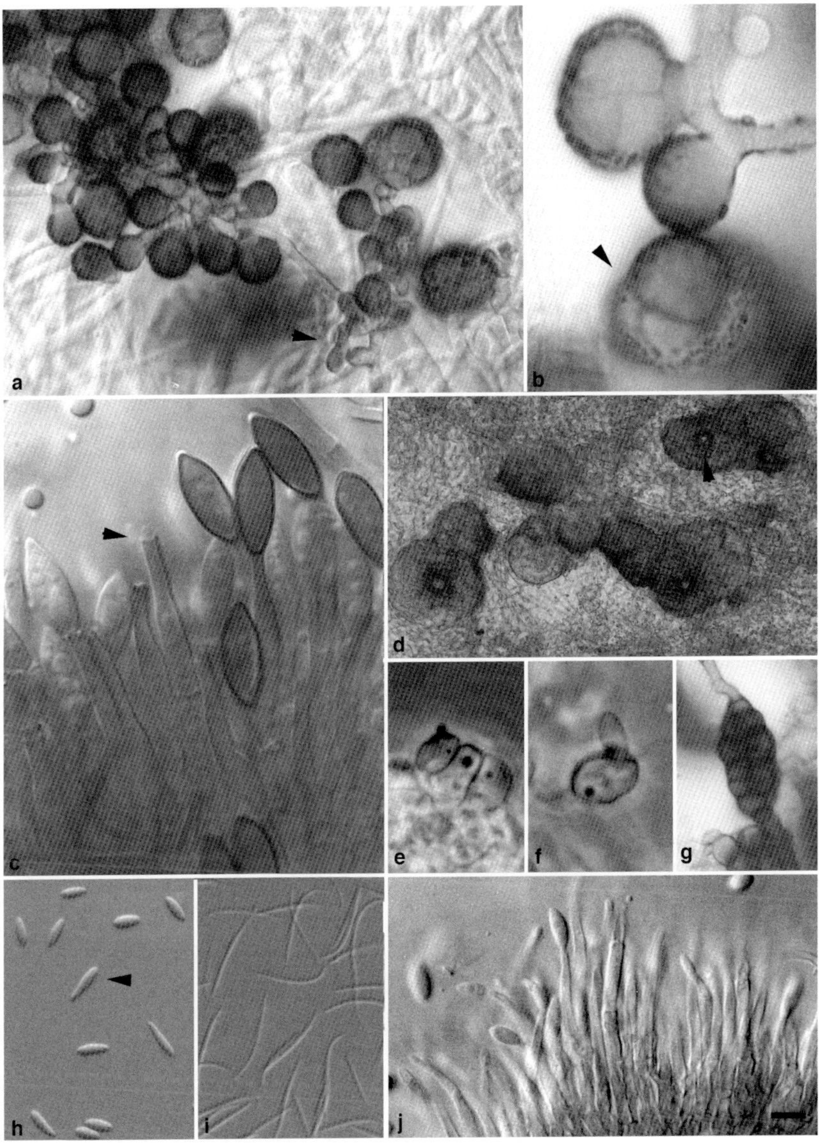

Fig. VIII.9. Coelomycetes

a, b. *Epicoccum nigrum*: **a.** Conidia formation in acervuli, young stages (arrow). Old conidia are septate, with transverse, longitudinal and oblique septa (muriform). **b.** Detail of holoblastic conidiogenesis, muriform conidia (arrow). **c.** *Melanconium apiocarpum*: Conidiophores in acervulus, basal tissue; arrow indicates site of conidiogenesis. **d–g.** *Phoma* sp.: **d.** Pycnidia with ostiolum (arrow). **e, f.** Conidiogenous cells (phialides). **g.** Multiseptate chlamydospore. **h–j.** *Phomopsis* sp.: **h.** Alpha conidia, intermediate conidia (arrow). **i.** Beta conidia. **j.** Acervulus with conidiophores. Scale: d: 40 μm; a, g–i: 10 μm; b, c, e, f, j: 4 μm.

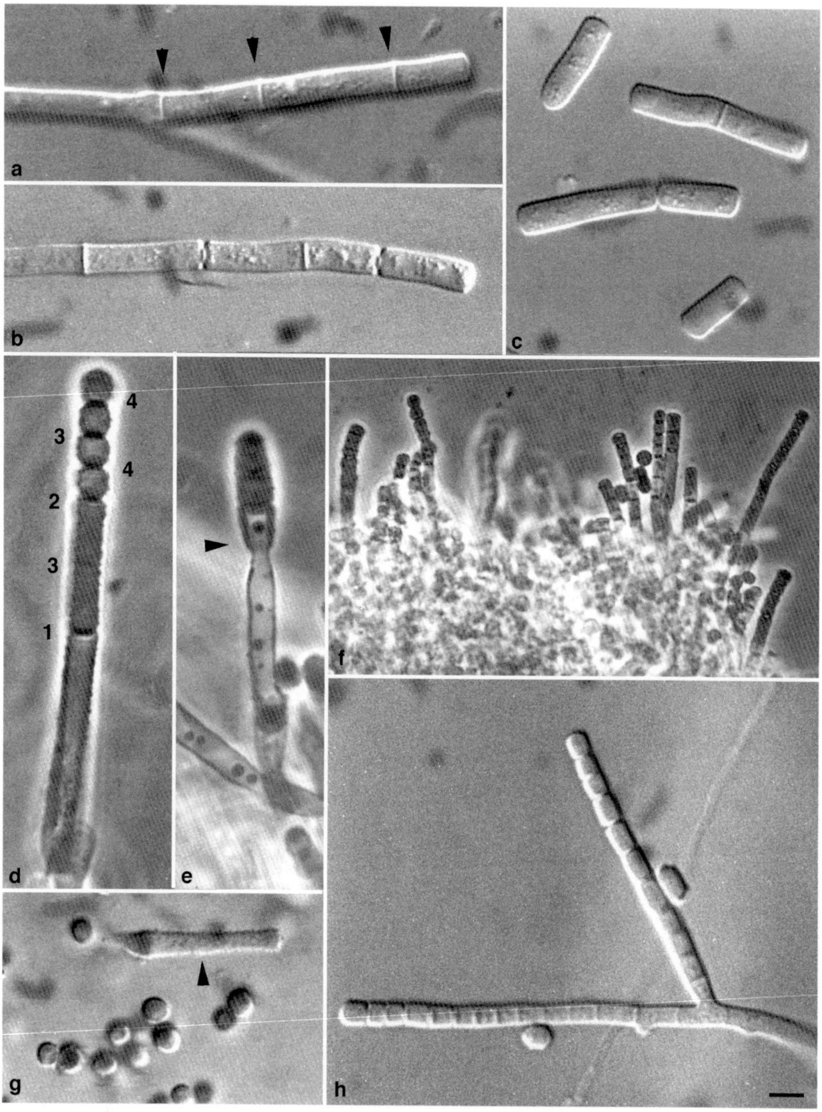

Fig. VIII.10. Thallic conidiogenesis

a–c. *Geotrichum candidum*: **a, b**. Septate hyphae (arrows) breaking up and forming conidia. **c**. Breaking hyphae and conidia. **d–g**. *Wallemia sebi*: **d**. Hypha with a first septum in the middle (1); additional septa are formed in the sequence 2–4, originating sequences of 4 conidia. **e**. Phialide-like cell (arrow), from which the conidiogenous hypha is growing. **f**. General view. **g**. Young conidia forming hypha (arrow); mature conidia, becoming spherical and warty when old. **h**. Basidiomycetous anamorph. Scale: a–e, g, h: 4 µm; f: 10 µm.

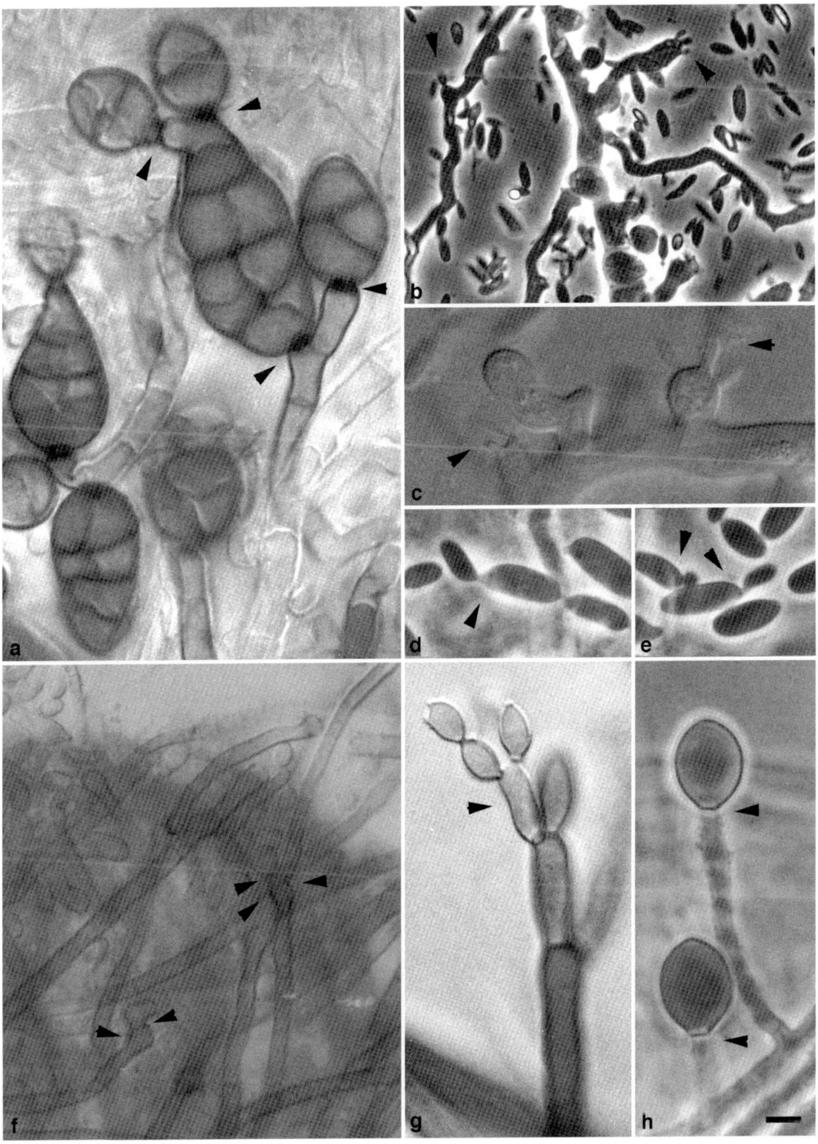

Fig. VIII.11. Holoblastic conidiogenesis – 1

a. *Alternaria alternata*: Poroconidia in acropetal chains on conidiophores; conidiogenous loci are melanised and in sympodial sequence (arrows); rostrate conidia with longitudinal, transverse and oblique septa. **b–e.** *Aureobasidium pullulans*: **b, c.** Conidiogenesis directly on hypha (arrows). **d, e.** Budding conidia (arrows). **f, g.** *Cladosporium cladosporioides*: **f.** Sympodially growing conidiophores (arrows: conidiogenous loci). **g.** Conidia in acropetal chains; conidium becoming a conidiogenous cell (arrow). **h.** *Humicola grisea*: blastoconidia with a broad base (arrow). Scale: b, f: 10 µm; a, c–e, g, h: 4 µm.

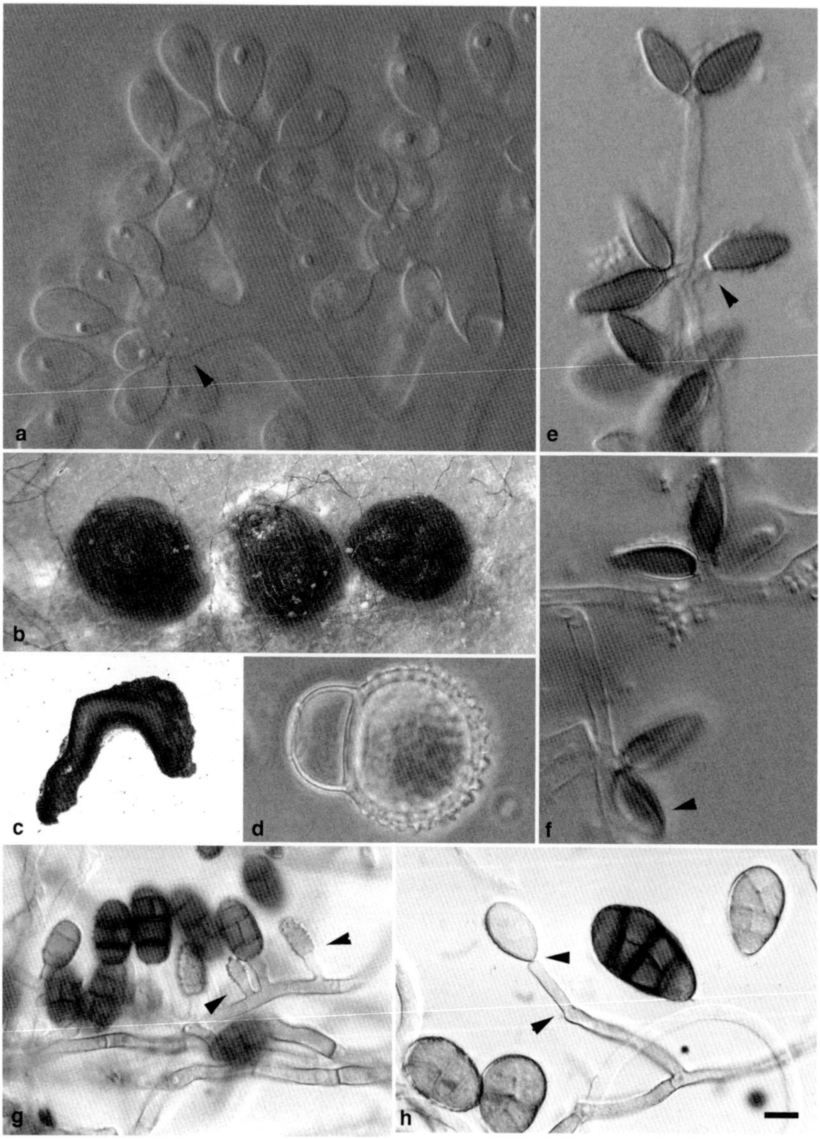

Fig. VIII.12. Holoblastic conidiogenesis – 2

a–c. *Botrytis cinerea*: **a.** Swollen conidiogenous cell (arrow) with simultaneous formation of conidia. **b.** Sclerotia. **c.** Section through a sclerotium showing the dark cortex and the lighter coloured medulla. **d.** *Mycogone rosea*: two-celled conidia. **e, f.** *Mammaria echinobotryoides*: blastoconidia with broad base (**e**, arrow) and germ slit (**f**, arrow). **g.** *Pythomyces chartarum*: muriform blastoconidia, with broad base; young stages (arrows). **h.** *Ulocladium chartarum*: conidiophore with sympodial conidiogenous loci (arrows); conidia without rostrum. Scale: b, c: 635 μm; g, h; 8 μm; a, d–f: 4 μm.

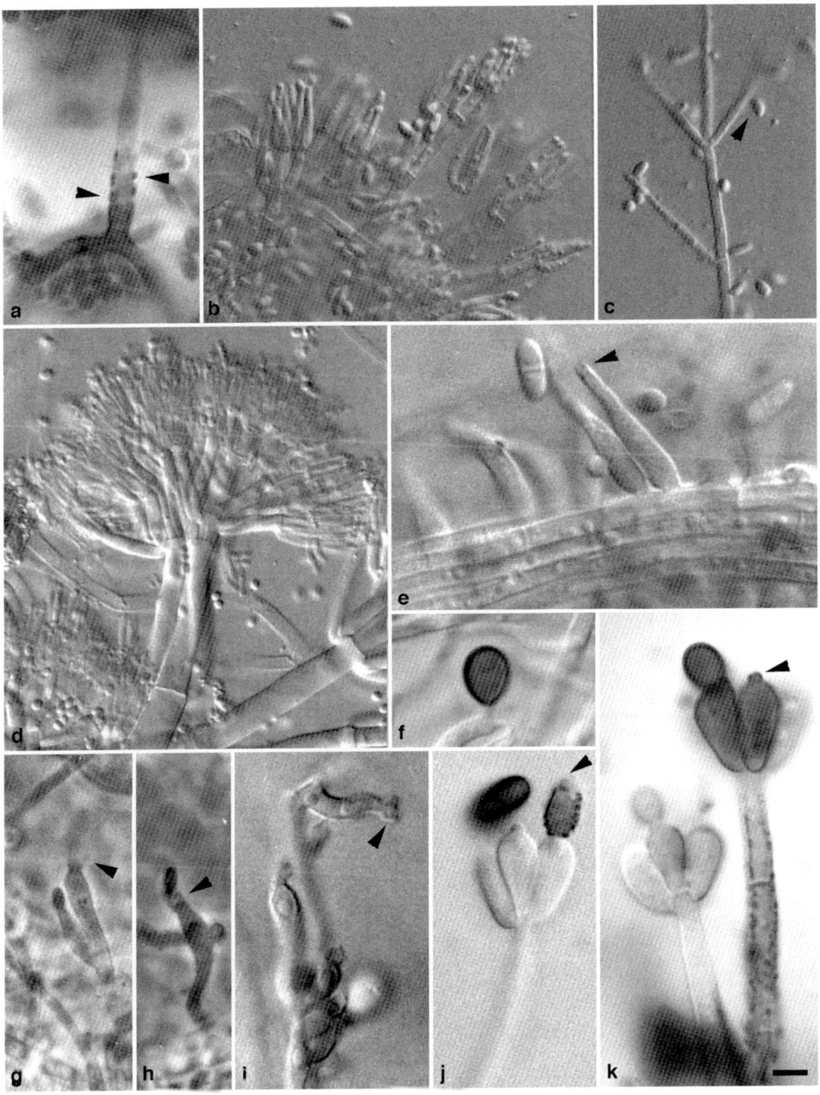

Fig. VIII.13. Enteroblastic conidiogenesis without chain formation – 1

a. *Acremonium rutilum*: phialide originating directly from hypha, with chromophilic base (arrows). **b, c.** *Clonostachys rosea*: **b.** Conidiophore with brush-like whorls of phialides. **c.** Conidiophore with verticillate phialides (*Verticillium*-type); asymmetric conidia (arrow). **d.** *Gliocladium viride*: brush-like branching conidiophore. **e.** *Monographella cucumerina*: phialide with collarette (arrow), formed directly on hypha; on the left two-celled conidium. **f–i.** *Phialophora mustea*: **f.** Aleurospore. **g–i.** Phialides with collarette (arrows). **j–k.** *Stachybotrys chartarum*: **j.** Old conidium with coarse cell wall (arrow). **k.** Bulbous phialides in whorls on the conidiophore tip, with protruding conidiogenous locus (arrow). Old conidiophores with coarse surface. Scale: b–d: 10 µm; a, e–k: 4 µm.

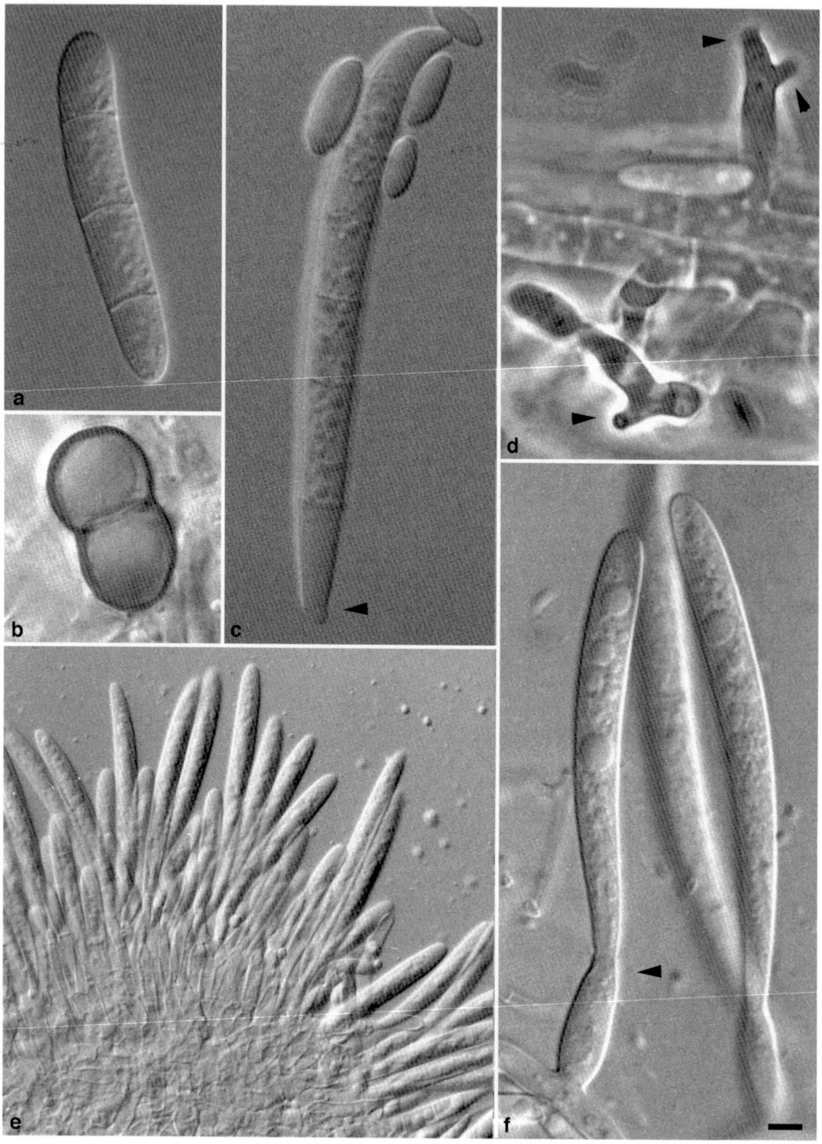

Fig. VIII.14. Enteroblastic conidiogenesis without chain formation – 2

a, b. *Cylindrocarpon destructans*: **a.** Multiseptate macroconidium with rounded spore ends. **b.** Chlamydospore. **c–f.** *Fusarium* spp.: **c.** Multiseptate macroconidium with foot cell (arrow); one-celled microconidia (top). **d.** Polyphialides (phialides with more than one conidiogenous locus, arrows). **e.** Sporodochium. **f.** Phialide (arrow). Scale: e: 10 µm; a–d, f: 4 µm.

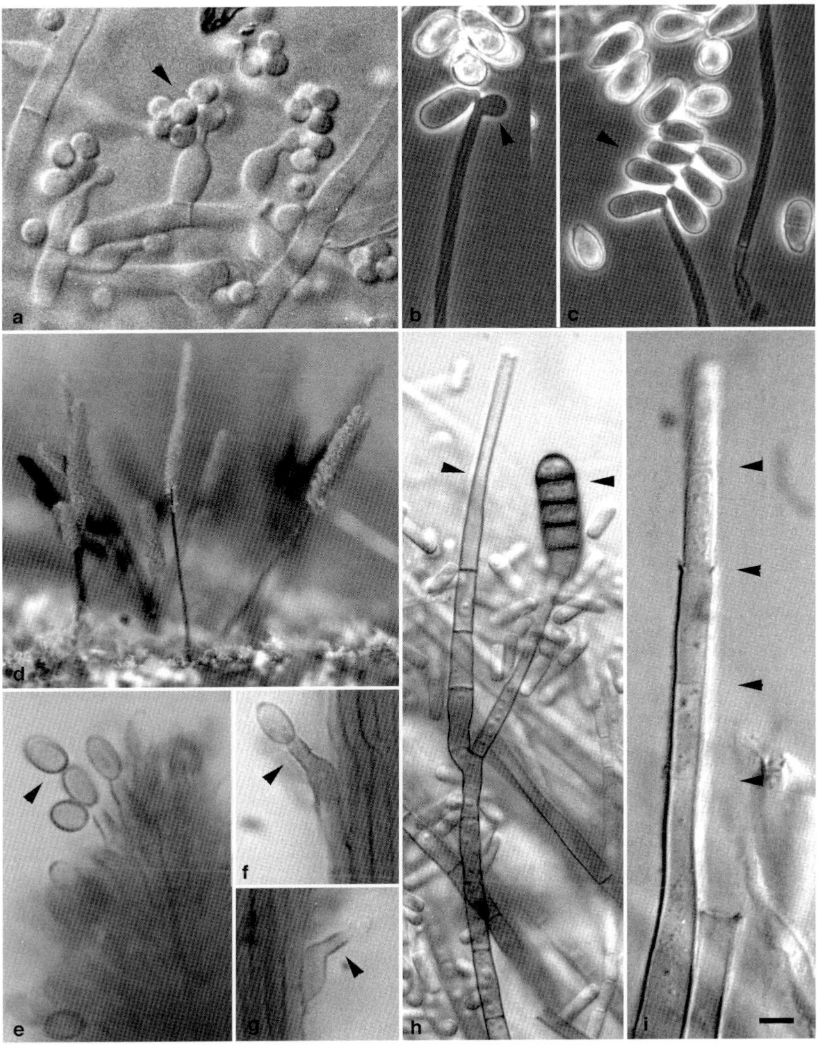

Fig. VIII.15. Enteroblastic conidiogenesis without chain formation (a), retrogressive (meristematic) conidiogenesis, conidia in chains (b, c), enteroblastic conidiogenesis with basipetal chain formation (d–i), anellides (d–g)

a. *Trichoderma* sp.: Phialides formed on poorly differentiated conidiophores, droplets of conidia (arrow). **b, c.** *Trichothecium roseum*: **b.** Conidium formation on a conidiophore (arrow); the previously formed conidium is still attached. **b.** Two-celled conidia in a chain (arrow). **d–g.** *Cephalotrichum microsporum*: **d.** Synnemata. **e.** Apex of a synnema with conidiogenous cells: chain of conidia (arrow). **f, g.** Conidiogenous cells (anellides) with rings (arrows). **h, i.** *Thielaviopsis basicola*: **h.** Conidiophore; phialide with long, cylindrical neck (left arrow: conidiogenous locus); four-celled arthroconidium, later disintegrating in one-celled propagules (right arrow). **i.** Conidial chain partly still within the phialide neck (arrows indicating septa where chain breaks). Scale: d: 200 µm; b, c, h: 10 µm; a, e–g, i: 4 µm.

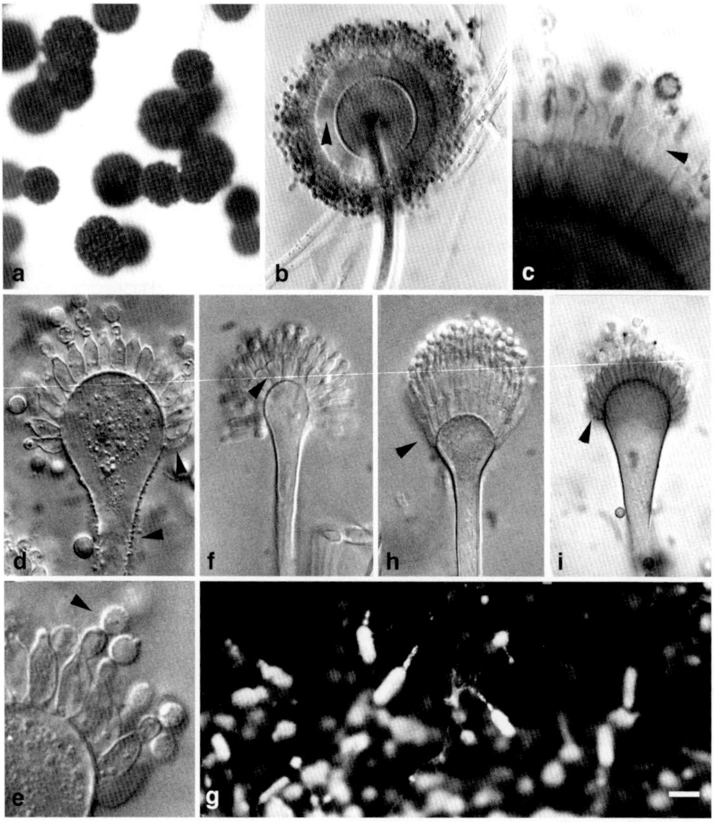

Fig. VIII.16. Enteroblastic conidiogenesis with chain formation: Aspergillus *spp. (see also Fig. IX.1)*

a–f. Phialides originating directly from the vesicle (uniseriate species) or from a row of sterile cells (metulae) (biseriate species). The phialides are disposed radially around the vesicle, thus giving rise to spherical heads. **a–c.** *A. niger* (biseriate species): **a.** Conidial head. **b.** Conidiophore with vesicle in the middle, on it metulae (arrow), phialides and conidia. **c.** Conidia, phialides (arrow) and metulae. **d, e.** *A. flavus* (uniseriate/biseriate species): **d.** Phialides (arrow) originating directly from the vesicle (uniseriate); coarse stipe (arrow); *A. flavus* may have biseriate heads as well, or a mix of both forms on the same head. **e.** Phialides, chain of conidia (arrow). **f.** *A. versicolor* (biseriate species): phialides on metulae (arrow), smooth stipe. **g–i.** Phialides only on the upper part of the vesicle, thus heads with a columnar appearance. **g, h.** *A. terreus* (biseriate species): **g.** Columns. **h.** Conidiophore with vesicle in the middle, metulae (arrow), phialides and conidia. **i.** *A. fumigatus* (uniseriate species): phialides (arrow) formed directly on the vesicle. Scale: a, g: 100 µm; b: 20 µm; d, f, h, i: 8 µm; c, e: 4 µm.

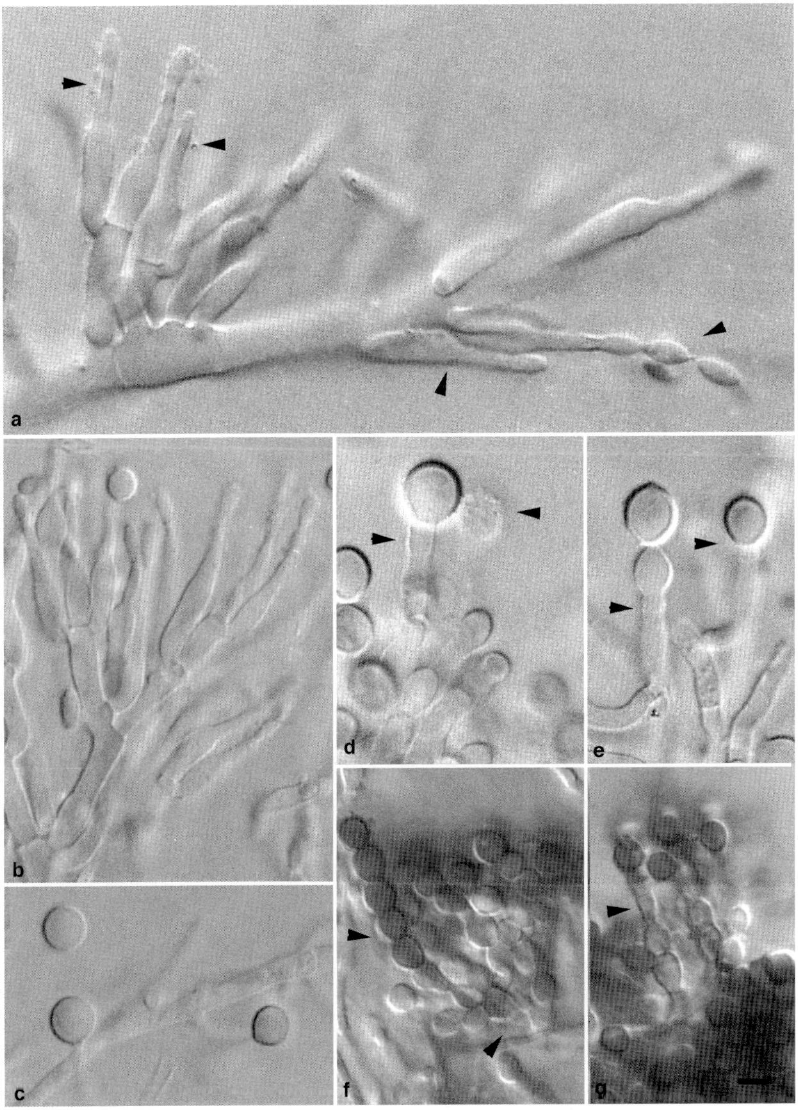

Fig. VIII.17. Enteroblastic conidiogenesis with chain formation (a–c), anellidic conidiogenesis (d–g)

a–c. *Paecilomyces variotii*: **a.** Tapering phialide (arrow in the middle); conidiogenous locus appearing dark (left arrows); conidia in chains (right arrow); **b.** Conidiophore with irregular branching. **c.** Aleurospores. **d, e.** *Scopulariopsis flava*: **d.** Conidiogenous cell (anellide) with rings (left arrow); old conidia with warty outer cell wall (right arrow). **e.** Anellides with rings (left arrow); conidia with truncate base (right arrow); conidia in chains (on left). **f, g.** *Scopulariopsis fusca*: **f.** Branched conidiophore (right arrow); conidia in chains (left arrow). **g.** Anellides with rings (arrow). Scale: a–g: 4 µm.

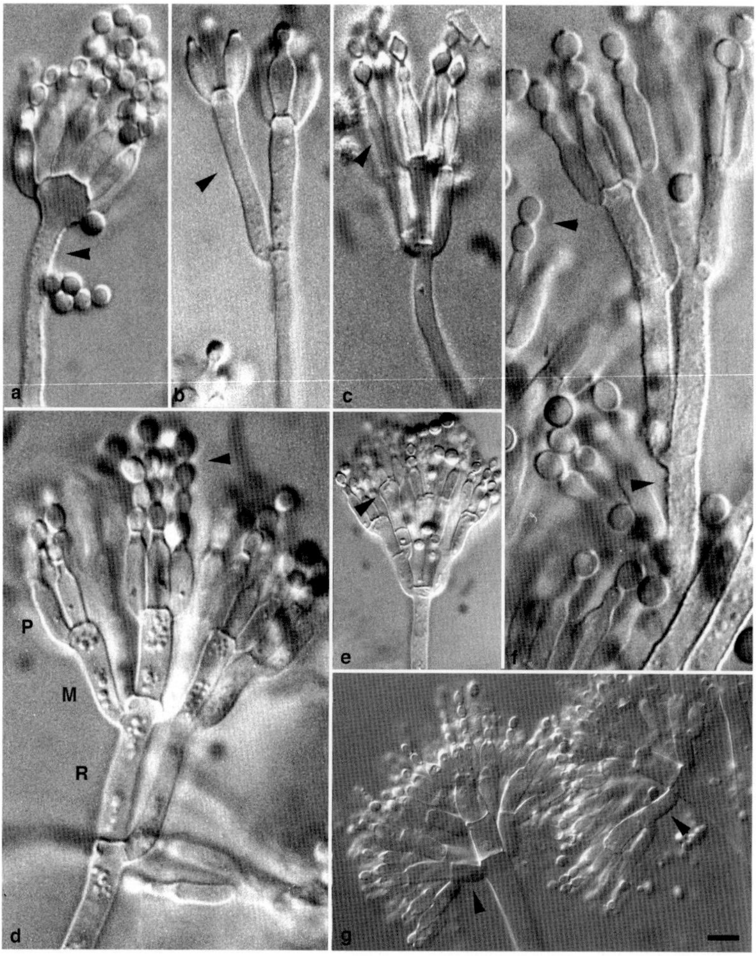

Fig. VIII.18. Enteroblastic conidiogenesis with chain formation:* Penicillium *spp. (see also Fig. IX.2)

a. *P. glabrum* (Subgenus *Aspergilloides*): phialides originating directly from a small swelling of the conidiophore, conidiophore with a coarse stipe (arrow). **b.** *P. corylophilum* (Subgenus *Furcatum*): metulae (arrow) of different lengths, longer than phialides. **c.** *P. variabile* (Subgenus *Biverticillium*): phialides (arrow) approximately as long as metulae. **d.** *P. chrysogenum* (Subgenus *Penicillium*): conidiophore with two orders of branches; phialide (P), metulae (M), rami (R); conidiophore stipe smooth; conidia in chains (arrow). **e.** *P. brevicompactum* (Subgenus *Penicillium*): penicilli appearing wider than high, through the broader metulae (arrow) and short rami. **f.** *P. commune* (Subgenus *Penicillium*): Conidiophore with a coarse stipe (lower arrow), conidia in chains (upper arrow). **g.** *P. olsonii* (Subgenus *Penicillium*): rami (arrows) and metulae very short and numerous. Scale: e, g: 8 µm; a–d, f: 4 µm.

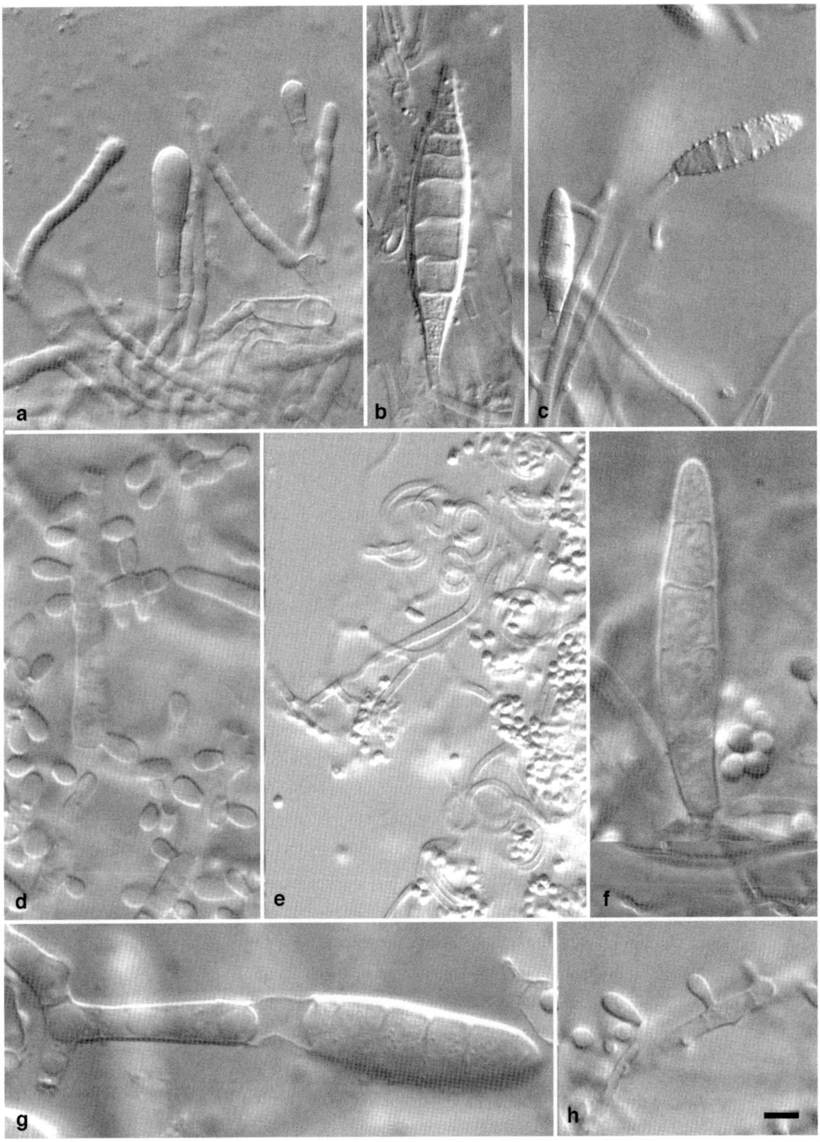

Fig. VIII.19. Dermatophytes – holothallic conidiogenesis

a. *Epidermophyton floccosum*: macrospore formation. **b.** *Microsporum canis*: rough-walled macrospore with thick cell wall. **c.** *Microsporum gypseum*: rough-walled macrospores with thin cell wall. **d.** *Trichophyton erinacei*. **e, f.** *Trichophyton mentagrophytes*: **e.** coiled hyphae and microspores. **f.** Macrospore. **g.** *Trichophyton ajelloi*: macrospore being detached. **h.** *Trichophyton rubrum*. Scale bar: a-c, e: 10 µm; d, g-h: 4 µm.

IX. Common mould genera

Information on the taxonomic position of the genera listed, as well as the number of taxa known, has been extracted from the CABI database (http://www.indexfungorum.org/). We have also consulted MycoBank (Crous et al. 2004; Robert et al. 2005) for additional data.

References in **boldface** are particularly useful for the identification of species. *Incertae sedis* means that the taxonomic position of a given genus is still controversial.

***Absidia* v. Tieghem** **Fig. VIII.2a–g**

***Mucoraceae*, Zygomycota**

NUMBER OF TAXA: 20

REFERENCES: **Domsch et al. (2007)**, Samson et al. (2004), Zycha et al. (1969)

EXAMPLES: *A. cylindrospora* Hagem, *A. spinosa* Lendner. Species of *Absidia* are mainly inhabiting soil and are frequently isolated from the rhizosphere or from rotten plant material.

***Acremonium* Link** **Fig. VIII.13a**

Anamorphs *incertae sedis* and *Hypocreales*, Ascomycota

NUMBER OF TAXA: approx. 80

REFERENCES: **De Hoog et al. (2000), Domsch et al. (2007), Gams (1971a)**, Peberdy (1987; biotechnology, secondary metabolites), Rossman et al. (1999), Samson et al. (2004), Summerbell et al. (2011; phylogeny).

EXAMPLES: *A. butyri* (v. Beyma) W. Gams: soil, leaves, wood, fungi. *A. charticola* (Lindau) W. Gams: i.a. on damp tapestry, walls, plants, air. *A. rutilum* W. Gams: soil. *A. strictum* W. Gams: soil, hay, leaves, fungi; sometimes as a human pathogen.

NOTES: Identification of species is often possible only for the specialist.

***Actinomucor* Schostak.**
Mucoraceae, Zygomycota

Number of taxa: 1

References: Benjamin & Hesseltine (1957), Domsch et al. (2007), Jong & Yuan (1985)

Example: *A. elegans* (Eidam) Benjamin & Hesseltine, the only species so far described, has been isolated form agricultural soil. The fungus is an excellent producer of proteases and is used in the fermentation of soybeans (Sufu or Tofu production).

***Alternaria* Nees** **Fig. VIII.11a**
Anamorphic Pleosporaceae, Ascomycota

Number of taxa: more than 100

Teleomorph: *Lewia* M.E. Barr & E.G. Simmons

References: **Ellis (1971, 1976)**, Rotem (1994; ecology), Samson et al. (2004), **Simmons** (1967, 1996a, 1996b, 1997, 1998, **2007**), Wiltshire (1947)

Example: *A. alternata* (Fr.) Keissler, (= *A. tenuis* Nees): cosmopolitan, secondary coloniser of dead plant material, sometimes causing rot of agricultural plants such as tomatoes. It may also grow on damp walls or tapestry and cause allergic asthma. It is used in allergenic tests.

***Aphanocladium* W. Gams**
Anamorphic *Nectriaceae*, Ascomycota

Number of taxa: 1

References: Gams (1971a)

Example: *Aphanocladium album* W. Gams: grows parasitically on button mushrooms.

Ascochyta Lib.
Anamorphs *incertae sedis* and *Pleosporales*, Ascomycota

NUMBER OF TAXA: more than 300

TELEOMORPH: *Didymella* Sacc.

REFERENCES: Boerema & Dorenbosch (1973), Buchanan (1987), Mel'nik (1977), **Punithalingam (1979, 1981, 1988)**, **Sutton (1980)**

EXAMPLES: *Ascochyta pisi* Lib., *Ascochyta cucumis* Fautr. & Roum.: leaf spots on beans and cucumber.

NOTE: Identification according to the host plant.

***Ascotricha* Berk.**
***Xylariaceae*, Ascomycota**

NUMBER OF TAXA: approx. 10

ANAMORPH: *Dicyma* Boulanger, *Nodulisporium* Preuss

REFERENCES: Ames (1963), Ellis (1971), Hawksworth (1971)

EXAMPLE: *A. chartarum* Berk.: on dead leaves and stems, on paper and cardboard, cosmopolitan.

***Aspergillus* P. Micheli ex Link** **Fig. VIII.16, IX.1**
Anamorphic *Trichocomaceae*, Ascomycota

NUMBER OF TAXA: more than 200

TELEOMORPH: *Dichlaena* Dur. & Mont., *Emericella* Berk. & Broome, *Eurotium* Link, *Fennellia* Wiley & Simmons, *Hemicarpenteles* Sarbhoy & Elphick, *Neocarpenteles* Udagawa & Uchiy., *Neopetromyces* Frisvad & Samson, *Neosartorya* Malloch & Cain, *Petromyces* Malloch & Cain

REFERENCES: **Domsch et al. (2007)**, Gravesen et al. (1994; detailed information on several species), **De Hoog et al. (2000**; medically relevant species), Frisvad & Samson (2000), Klich (1993, **2002**), **Kozakiewicz (1989)**, Mücke & Lemmen (2005; clinical aspects, mycotoxins), Pitt & Samson (1993), Powell et al. (1994;

industrial use), **Raper & Fennel (1965)**, Samson (1979, 1992, 1994), Samson & Gams (1985), Samson & Pitt (1985, 1990, 2000; molecular taxonomy), **Samson et al. (2004)**, **Samson & Varga (2007)**, **Samson et al. (2011)**, Singh et al. (1991; in particular species on cereals).

Examples: *A. flavus* Link ex Gray and *A. parasiticus* Speare: on nuts and cereals, both producers of aflatoxins. *A. fumigatus* Fres.: thermotolerant, found in hay, compost, organic waste, house dust, air; human pathogen, causing lung aspergillosis. *A. niger* v. Tiegh.: cosmopolitan, causing spoilage of food and materials such as paper, leather and clothes; used for the production of organic acids, in particular citric acid. *A. terreus* Thom: in soil, the rhizosphere, clothes, air, house dust; used for mevinolin synthesis, a compound used to lower blood cholesterol levels.

Note: The teleomorphs *Chaetosartorya* und *Sclerocleista* are now included in *Emericella* and *Hemicarpenteles, Edyuillia* and *Warcupiella* in *Emericella*.

Morphologically, species of *Aspergillus* can be distinguished by the position of the conidiogenous cells on the vesicles (Fig. IX.1) as well as by cultural characters. Taxonomic groups are formed based on their teleomorphs and secondary metabolite production. Many *Aspergillus* species are soil colonisers and xerophilic. Several species may cause economically important food and storage spoilage (e.g. stored cereals, coffee beans); they also are important toxin producers and may sometimes cause clinically severe allergies (e.g. *A. fumigatus* spores and volatile products). Some species (e.g. *A. flavus*, *A. fumigatus, A. niger*, *A. terreus*) are opportunistic human pathogens. Other species are used in the food industry for the production of fermented food such as soy sauce, vitamins, enzymes, antibiotics and organic acids. Stored cereal seeds colonised by *Aspergillus* have a reduced germination rate, the embryo is pigmented and the fatty acid content is modified.

For human and animal pathogenic Aspergilli, see the very informative, dedicated website (http://www.aspergillus.org.uk/).

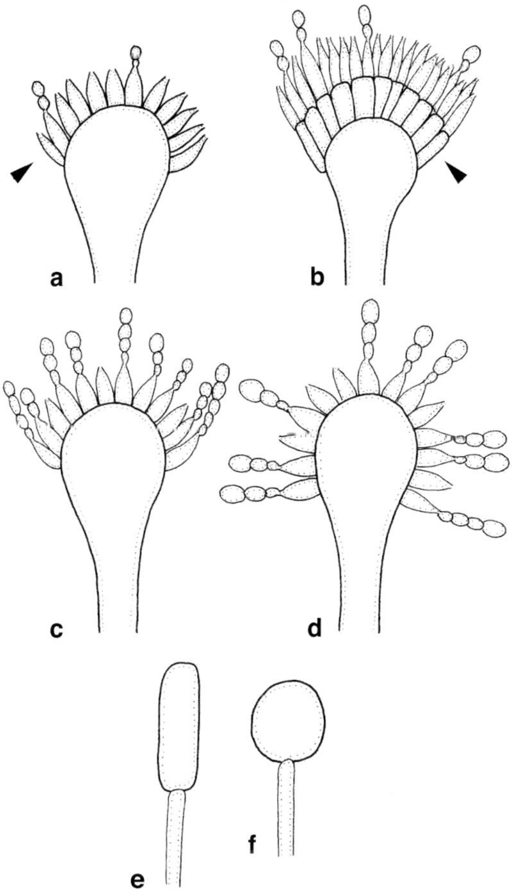

Fig. IX.1. Subgeneric classification of **Aspergillus**

a. Phialides (arrow) formed directly on the vesicle. **b.** Phialides formed on a row of sterile cells (metulae; arrow). **c.** Phialides situated only on the superior part of the vesicle; in this case, the conidial chains have the appearance of a column. **d.** Phialides are formed all around the vesicle, giving the structure a spherical appearance. **e.** Columnar head. **f.** Spherical head.

Aureobasidium Viala & Boyer — Fig. VIII.11b–e

Anamorphic *Dothioraceae*, Ascomycota

NUMBER OF TAXA: approx. 16

TELEOMORPH: *Discosphaerina*

References:	De Hoog (1999), De Hoog & Hermanides-Nijhof (1977), De Hoog & Grube (2008; black yeasts), **Hermanides-Nijhof (1977)**, Samson et al. (2004), Zalar et al. (2008)
Example:	*A. pullulans* (De Bary) Arnaud: saprobe on cereal seeds, flour, nuts, tomatoes, fruit and fruit juices; frozen fruit pies; on skin and nails and occasionally also on damp walls, windows, silicon joints, sometimes causing a paint blackening. It can survive fungicides contained in the paint.
Notes:	Conidiogenesis in *Aureobasidium* is synchronous, as opposed to *Hormonema*. A number of genera such as *Exophiala*, *Hormonema*, *Moniliella*, *Ramichloridium* und *Rhinocladiella*, the so-called **black yeasts**, can be easily confused with *Aureobasidium* by the beginner (and often by advanced mycologist). For the experienced mycologists as well, however, a reliable identification is often possible only using molecular biology techniques.

Basipetospora **G.T. Cole & W.B. Kendr.** — **Fig. VIII.6g–i**
Anamorphic *Monascaceae*, Ascomycota

Number of taxa:	3
Teleomorph:	*Monascus*
Example:	*Basipetospora chlamydospora* Matsush.

Bipolaris **Shoemaker**
Anamorphic *Pleosporaceae*, Ascomycota

Number of taxa:	approx. 30
Teleomorph:	*Cochliobolus* Drechsler
References:	Ellis (1971, 1976), Sivanesan (1987), Smiley et al. (2000; pathogenic species of turfgrass)
Example:	*B. maydis* (Nisikado & Miyake) Shoemaker: leaf spot disease of corn in the southern hemisphere.
Notes:	In some identification keys *Bipolaris* is included in *Drechslera*. Species of *Bipolaris* live mostly saprobically on *Poaceae*. Selected species may cause blight symptoms on turfgrass.

Botrytis P. Micheli ex Pers. **Fig. VIII.12a–c**
Anamorphic _Sclerotiniaceae_, Ascomycota

NUMBER OF TAXA: approx. 30

TELEOMORPH: i.a. *Botryotinia* Whetzel (= *Sclerotinia* Fuckel p.p.)

REFERENCES: Domsch et al. (2007), **Ellis (1971)**, Jarvis (1977), Samson et al. (2004)

EXAMPLES: *B. aclada* Fres. (= *B. allii* Munn): on *Alllium*, neck rot disease in onions during storage. *B. cinerea* Pers.: plant pathogen, the so-called grey mould on stored plant products, in particular strawberries and vegetables; It is also the cause of the "noble rot": colonised grapes are used in France to produce an excellent sweet wine. It is also a good producer of pectinases used in the industrial fruit juice production. *B. cinerea* is a species complex and, correspondingly, the anamorph of several ascomycetous genera. The spore release is dependent on humidity changes.

Byssochlamys Westling **Fig. VIII.5a–d**
Trichocomaceae, Ascomycota

NUMBER OF TAXA: 4

ANAMORPH: *Paecilomyces*

REFERENCES: **Samson (1974), Samson et al. (2004)**, Stolk & Samson (1971)

EXAMPLES: *B. fulva* Olliver & G. Smith: in fruit juices and preserves, on stored grapes, in the soil. *B. nivea* Westling: in soil, on cereals, damaged fruit, fruit juice.

NOTES: Ascospores are heat-resistant and germinate after a heat shock.

Cephalotrichum Link **Fig. VIII.15d–g**
Anamorphic _Microascaceae_, Ascomycota

SYNONYM: *Doratomyces* Corda

NUMBER OF TAXA: 4

References:	Morton & Smith (1963), Ellis (1971), Domsch et al. (2007)
Example:	*C. stemonitis* (Pers. ex Steud.) Morton & G. Smith: on wood, grass stems, dung, in soil and compost.

Ceuthospora Gréville
Anamorphic _Phacidiaceae_, Ascomycota

Number of taxa:	more than 100
Teleomorph:	*Phacidium* Fr.
References:	DiCosmo et al. (1983, 1984), **Nag Raj (1993)**, Sutton (1980)
Example:	*C. pinastri* (Fr.) Höhnel on pine needles.

Chaetomium Kunze **Fig. VIII.7d–f**
Chaetomiaceae, Ascomycota

Number of taxa:	approx. 100
Anamorph i.a.:	*Botryotrichum* Sacc. & March.
References:	Ames (1963), von Arx et al. (1984, 1986), **Domsch et al. (2007), Dreyfuss (1975)**
Example:	*C. globosum* Kunze: very common on cellulose-containing substrates (cardboard, old paper, plant remains), in soil and compost.

Chrysonilia v. Arx
Anamorphic _Sordariaceae_, Ascomycota

Number of taxa:	3
Teleomorph:	*Neurospora*
References:	von Arx (1981a), Samson et al. (2004)
Example:	*C. sitophila* (Mont.) von Arx; very rapid growth; commonly found on bread ("red mould" on bread – not to be confused with the "red bread mould" decay caused by the bacterium

Serratia marcescens), and meat; it may also cause rot of strawberry, raspberry and apple; it is a feared lab contamination because of its extremely rapid growth and abundant conidia production.

Chrysosporium Corda
Anamorphic *Onygenaceae*, Ascomycota

NUMBER OF TAXA: approx. 30

TELEOMORPH: *Aphanoascus* Zukal, *Apinisia* La Touche, *Arthroderma* Currey, *Bettsia* Skou, *Corynascus* v. Arx, *Gymnoascus* Baran., *Pectinotrichum* Varsausky & Orr, *Pseudogymnoascus* Raillo, *Renispora* Sigler & J. Carm., *Rollandina* Pat.

REFERENCES: Carmichael (1962), **Sigler & Carmichael (1976)**, van Oorschot (1980)

EXAMPLES: *C. farinicola* (Burnside) Skou (= *C. fastidium* Pitt), *C. inops* Carm., *C. xerophilum* Pitt: xerophilic, on foodstuff with a low water activity.

Cladosporium Link
Fig. VIII.11f–g

Anamorphic *Davidiellaceae*, Ascomycota

NUMBER OF TAXA: approx. 35

TELEOMORPH: *Davidiella* Crous & U. Braun

REFERENCES: **Bensch et al. (2010, 2012)**, Crous et al. (2007), David (1997), Domsch et al. (2007), Ellis (1971), Ho et al. (1999), Samson et al. (2004), Schubert et al. (2007), Zalar et al. (2007)

EXAMPLES: *C. cladosporioides* (Fres.) de Vries, *C. herbarum* (Pers.) Link, *C. macrocarpum* Preuss, *C. sphaerospermum* Penzig: cosmopolitan, in soil, on foodstuff, cereals, in the air. *C. cucumerinum* Ellis & Arth.: scab of cucurbits and pumpkin, squash and marrow.

NOTES: Species of *Cladosporium* grow at high water activity (a_w>0.85). In particular *C. cladosporioides* and *C. herbarum* may form prominent black spots and films in damp rooms (bathroom, showers), on window frames and silicon seals. Some strains may grow at very low temperatures (–6 °C). *Cladosporium* species may also cause black discolorations of clothing and tapestry. The fungus is also used (as *Alternaria*) for allergenic tests.

Clonostachys Corda **Fig. VIII.13b, c**
Anamorphic _Bionectriaceae_, Ascomycota

NUMBER OF TAXA: 10

TELEOMORPH: *Bionectria* Speg.

REFERENCES: Rossman et al. (1999), Schroers (2001), Schroers et al. (1999)

EXAMPLE: *Cl. rosea* (Link : Fr.) Schroers et al. [Syn.: *Gliocladium roseum* (Link) Thom; Tel.: *Bionectria ochroleuca* (Schw.) Schroers et al. [Syn.: *Nectria ochroleuca* (Schw.) Berk.]: soil fungus, colonising dead, rotten plant material. It is also used as biocontrol agent because it can grow on other fungi as a hyperparasite.

Colletotrichum Corda **Fig. VIII.8d, e**
Anamorphic _Glomerellaceae_, Ascomycota

NUMBER OF TAXA: approx. 40

TELEOMORPH: *Glomerella* Spauld. & Schrenk

REFERENCES: von Arx (1957, 1970), Damm et al. (2012), Prusky et al. (2000; plant pathological issues), **Sutton (1980)**

EXAMPLES: *C. acutatum* Simmonds: strawberry anthracnose. *C. gloeosporioides* (Penz.) Sacc. (Syn.: *Gloeosporium fructigenum* Berk.): causing anthracnose on several crops. *C. trichellum* (Fr. : Fr.) Duke: anthracnose of ivy.

Coniochaeta (Sacc.) Cooke
Coniochaetaceae, Ascomycota

NUMBER OF TAXA: 35

ANAMORPH: *Lecythophora* Nannf.

References: Checa et al. (1988), **Hawksworth & Yip (1981)**, Weber (2002)

Example: *Coniochaeta ligniaria* (Grév.) Massee.

Note: Most species live saprobically on wood.

Cryptosporiopsis Bubák & Kabát **Fig. VIII.8f, g**
Anamorphic _Dermateaceae_, Ascomycota

Number of taxa: approx. 20

Teleomorph: *Pezicula* Tul. & Tul.

References: Sutton (1980), **Verkley (1999)**

Example: *C. malicorticis* (Cordl.) Nannfeldt (Syn.: *Gloeosporium malicorticis* Cordl.; *Gloesporium perennans* Zeller & Childs): Gloeosporium storage rot of apple and pears.

Cunninghamella Matr. **Fig. VIII.2i–j**
Cunninghamellaceae, Zygomycota

Number of taxa: 9

References: Domsch et al. (2007), Samson (1969), **Zheng & Chen (2001)**

Example: *C. elegans* Lendner: generally isolated from soil, rarely from foodstuff.

Curvularia Boedjin
Anamorphic _Pleosporaceae_, Ascomycota

Number of taxa: more than 50

Teleomorph: *Cochliobolus* Drechsler

References: Ellis (1966, 1971, 1976), Sivanesan (1987), Smiley et al. (2000; pathogenic species on turfgrass), Watanabe (2002)

Note: Most species live saprobically, only a few may cause damages to turfgrass. Human pathogen in immunocompromised patients.

***Cylindrocarpon* Wollenw.** **Abb. VIII.14a, b**

Anamorphic *Nectriaceae*, Ascomycota

Number of taxa: approx. 30

Teleomorph: *Neonectria* Wollenw.

References: **Booth (1959, 1966)**, Domsch et al. (2007), Gerlach (1959a, 1959b), Rossman et al. (1999)

Example: *C. destructans* (Zins.) Scholten: soil fungus, sometimes isolated from roots.

***Cylindrocladium* Morgan**

Anamorphic *Nectriaceae*, Ascomycota

Number of taxa: approx. 20

Teleomorph: *Calonectria* de Not.

References: Boedijn & Reitsma (1950), **Crous (2002**; taxonomy, pathology, molecular biology), **Crous & Wingfield (1994)**, **Lombard et al. (2010)**, **Peerally (1991)**, Rossman et al. (1999)

Example: *C. scoparium* Morg.: root pathogen, isolated from soil.

Notes: *Cylindrocladium* conidiophores are branched apically and form a long, prominent stipe extension that terminates in a club-shaped vesicle.

***Dichlaena* Durieu & Mont.**

***Trichocomaceae*, Ascomycota**

Number of taxa: 1–2

Anamorph: *Aspergillus glaucus*-Group

References: Malloch & Cain (1972), Samson (1979)

Example: *D. lentisci* Mont. & Dur.: yellow-orange, hard stromata.

Didymella **Sacc.** **Fig. VIII.8a**

Anamorphs *incertae sedis* and *Pleosporales*, Ascomycota

NUMBER OF TAXA: approx. 75

ANAMORPH: *Ascochyta*, *Phoma*

REFERENCES: Corbaz (1957), Corlett (1981), Punithalingam (1981), Sivanesan (1984)

EXAMPLES: *D. bryoniae* (Fuckel) Rehm [= *Mycosphaerella citrullina* (C.O. Sm.) Grossenb.]: cucumber rot; *D. applanata* (Niessl.) Sacc.: on raspberry shrubs.

Diplodia **Fr.**

Anamorphic *Botryosphaeriaceae*, Ascomycota

NUMBER OF TAXA: 24 valid names, but the genus is in need of a comprehensive taxonomic revision.

REFERENCES: Sutton (1980), Zambettakis (1954)

EXAMPLE: *Diplodia malorum* Fuckel.

NOTE: Overall a taxonomically insufficiently known genus.

Discula **Sacc.** **Fig. VIII.8h–i**

Anamorphic *Gnomoniaceae*, Ascomycota

NUMBER OF TAXA: 10

TELEOMORPH: *Apiognomonia* von Höhnel

REFERENCES: Monod (1983), von Arx (1957, 1970), Hämmerli et al. (1992)

EXAMPLES: *D. umbrinella* (Berk. &. Br.) M. Morelet [teleomorph: *Apiognomonia errabunda* (Roberge ex Desm.) Höhn.], *D. platani* (Peck) Sacc. [teleomorph: *A. veneta* (Sacc. & Speg.) Höhn.]: Leaf spots (anthracnose) on their hosts.

***Drechslera* Ito**
Anamorphic *Pleosporaceae*, Ascomycota

NUMBER OF TAXA: more than 20

TELEOMORPH: *Pyrenophora* Fr.

REFERENCES: Ellis (1971, 1976), Sivanesan (1987), Smiley et al. (2000; pathogens of turfgrass)

EXAMPLE: *D. dactylidis* Shoemaker: saprobe of *Dactylis glomerata.*

NOTES: Most species live as saprobes on Poaceae; a few species may cause leaf spots on grasses, particularly turfgrass. Human pathogen in immunocompromised patients.

***Emericella* Berk.** **Fig. VIII.5e–j**
***Trichocomaceae*, Ascomycota**

NUMBER OF TAXA: more than 50

ANAMORPH: *Aspergillus* Sect. *Nidulantes*

REFERENCES: Domsch et al. (2007), Raper & Fennell (1965), Samson (1979), Samson et al. (2004)

EXAMPLE: *E. nidulans* (Eidam) Vuill.: soil fungus, isolated also from air and potatoes, seeds, cereals, cotton; it may also infect humans and animals (opportunistic infections). Known to cause aspergillosis in birds. Used for genetic studies in the laboratory.

***Epicoccum* Link** **Fig. VIII.9a, b**
Anamorphic *Pleosporaceae*, Ascomycota

NUMBER OF TAXA: 14

REFERENCES: Domsch et al. (2007), Samson et al. (2004), Sutton (1980)

EXAMPLE: *E. nigrum* Link (= *E. purpurascens* Ehrenb.): secondary coloniser of dead plant material; isolated from soil, seeds and beans. Endophyte of several host plants.

***Epidermophyton* E. Lang** **Fig. VIII.19a**
Anamorphic *Arthrodermataceae*

NUMBER OF TAXA: 1

TELEOMORPH: *Arthroderma* Curr.

REFERENCES: **de Hoog et al. (2000)**, Gräser et al. (2008), **St. Germain & Summerbell (2010)**

EXAMPLE: *E. floccosum* (Harz) Langer. et Milochevitch: human pathogen, causing tinea pedis, tinea cruris, tinea corporis and sometimes onychomycoses.

NOTE: *Epidermophyton* can be distinguished from *Trichophyton* and *Microsporum* by the absence of microconidia and the relatively thin-walled macroconidia, which develop as lateral or apical outgrowths of hyphae. The second species, *E. stockdaleae* Prochacki & Engelhardt-Zasada, has now been transferred to *Trichophyton ajelloi* (Vanbreuseghem) Ajello. It grows on 7% NaCl.

***Eupenicillium* Ludwig**
***Trichocomaceae*, Ascomycota**

NUMBER OF TAXA: approx. 10; a large number of anamorphic forms are described

ANAMORPH: *Penicillium*

REFERENCES: **Domsch et al. (2007)**, **Pitt** (1974, 1979, **2000**), Stolk & Samson (1983)

EXAMPLE: *E. lapidosum* Scott & Stolk: isolated from several soil types, including the wheat rhizosphere; from leaves, frozen fruit pies, canned blueberries, corn grains, rotting oranges. The ascospores are exceptionally heat resistant (terminal killing temperature: >70 °C during 30 min.); able to live under anaerobic (microaerobic) conditions. The valid name for *Hamigera striata* (Raper & Fennell) Stolk & Samson is *Penicillium lineatum* Pitt and this species should be included in *Eupenicillium.* This fungus is often found in fruit preserves and canned fruit and has also very heat-resistant ascospores.

NOTE: Many species produce the toxic compound patulin. According to Index fungorum (http://www.indexfungorum.org/) *Hamigera* and (partly) *Talaromyces* species are now included in *Eupenicillium.*

Eurotium Link

Fig. VIII.6a–e

***Trichocomaceae*, Ascomycota**

Number of taxa: more than 20

Anamorph: *Aspergillus* Sect. *Aspergillus*

References: **Blaser (1976)**, **Kozakiewicz (1989)**, Samson et al. (2004), Splittstoesser et al. (1989; heat-resistant species)

Examples: *E. amstelodami* Mangin, *E. chevalieri* Mangin, *E. herbariorum* (Wiggers) Link: cosmopolitan, mainly in tropical and subtropical regions; in soil, grains, fruit and fruit juices, peas, rice, dry foodstuff (xerophilic species); opportunistic human pathogens.

Note: *Eurotium* species may produce heat-resistant ascospores that germinate after a heat shock (30–60 min. at 60 °C).

Exophiala Carmichael

Anamorphic *Herpotrichiellaceae*, Ascomycota

Number of taxa: ca. 20

Teleomorph: *Capronia* Sacc.

References: **De Hoog (1977**, 1999), Domsch et al. (2007), Rogers et al. (1999)

Note: See discussion under *Aureobasidium*.

Exserohilum Leonard & Suggs

Anamorphic *Pleosporaceae*, Ascomycota

Number of taxa: approx. 11 valid names, but the taxonomy of the genus needs a revision (more species described).

Teleomorph: *Setosphaeria* Leonard & Suggs

References: Ellis (1971, 1976, as *Drechslera*), **Sivanesan (1987)**, **Smiley et al. (2000**; pathogenic species on grasses)

Example: *E. turcicum* (Pass.) Leonard & Suggs: worldwide distribution, on millet and corn; the causal agent of northern leaf blight in corn.

NOTE: *Exserohilum* species are mainly saprobes or pathogens on Poaceae, often on turfgrass.

Fennellia Wiley & Simmons
Trichocomaceae, Ascomycota

NUMBER OF TAXA: 2

ANAMORPH: *Aspergillus*

REFERENCES: Domsch et al. (2007), Samson (1979)

EXAMPLES: *F. nivea* (Wiley & Simmons) Samson: cosmopolitan, found mainly in tropical, subtropical and Mediterranean regions; inhabiting different soil types, rhizosphere; halophilic, toxin producer.

Fraseriella Cif. & A.M. Corte
Anamorphic _Monascaceae_, Ascomycota

NUMBER OF TAXA: 1

TELEOMORPH: *Xeromyces*

REFERENCES: Samson et al. (2004)

EXAMPLE: *Fraseriella bispora* Cif. & A.M. Corte

NOTE: See also *Xeromyces.*

Fusarium Link **Fig. VIII.14c–f**
Anamorphic _Nectriaceae_, Ascomycota

NUMBER OF TAXA: more than 40

TELEOMORPH: *Gibberella* Sacc.

REFERENCES: Booth (1959, 1971b, 1977), Burgess et al. (1988; laboratory guide), Chelkowski (1989), Desjardin (2006; toxins), Domsch et al. (2007), **Gerlach & Nirenberg (1982)**, Joffe (1986), **Leslie & Summerell (2006)**, Marasas et al. (1984; toxins), Mücke

& Lemmen (2004; toxins), **Nelson et al. (1983)**, Rossman et al. (1999), **Samson et al. (2004)**, Sing et al. (1991), Summerell et al. (2001)

EXAMPLES: *F. graminearum* Schwabe: cosmopolitan, on Poaceae, particularly corn and barley; toxin producer (trichothecenes, zearalenone).
F. oxysporum Schlecht.: Fr.: cosmopolitan, sometimes plant pathogen, often saprobe in soil. Endophyte of cereal seeds, fruit and vegetables.
F. venenatum Nirenberg: production of mycoproteins for human consumption (Quorn®).

NOTES: The identification of *Fusarium* relies on the morphology of the conidiogenous cells (monophialides, polyphialides) and conidia (macroconidia, microconidia). Secondary metabolite production and vegetative compatibility are often used as complementary features; the use of molecular biology techniques, however, is often crucial for an accurate identification. Leslie & Summerell (2006) provide a thorough description of the known species and useful information on the most reliable identification methods. Most *Fusarium* species are cosmopolitan and soil-borne. The genus includes plant pathogens known as the causal agents of different diseases, mostly wilt or blight, on vegetables, cereals and fruit. Some species may cause opportunistic infections in humans, while others are feared spoilage fungi. Several of them are toxin producers (i.a. trichothecenes, fumonisins, zearalenone).

Geotrichum Link — Fig. VIII.10a–c

Anamorphic *Dipodascaceae*, Ascomycota

NUMBER OF TAXA: 23

TELEOMORPH: *Dipodascus* Agerh., *Galactomyces* Redhead & Malloch

REFERENCES: **De Hoog et al. (1986)**, Domsch et al. (2007), Samson et al. (2004), Sigler & Carmichael (1976)

EXAMPLE: *G. candidum* Link [= *Oidium lactis* Fres., Tel.: *Galactomyces geotrichum* (Butler & Petersen) Redhead & Malloch]: cosmopolitan, sometimes causing opportunistic mycoses in humans and animals; commonly on nutrient-rich, humid substrates and in polluted water. Causing spoilage of fruit, vegetable, and dairy products; the fungus is able to grow in a microaerobic environment and sometimes even anaerobically. Cultures are yeast-like but the fungus forms a true mycelium.

***Gelasinospora* Dowding**
***Sordariaceae*, Ascomycota**

NUMBER OF TAXA: 20

REFERENCES: Lundqvist (1972), von Arx (1982)

EXAMPLE: *G. reticulispora* (Greis & Greis-Dengler) C. & M. Moreau: on dead plant remains.

***Gibberella* Sacc.**
***Nectriaceae*, Ascomycota**

NUMBER OF TAXA: ca. 20

ANAMORPH: *Fusarium* Link

REFERENCES: Booth (1971b), Rossman et al. (1999)

EXAMPLE: *G. fujikuroi* (Sawada) Wollenw. (Anam.: *F. fujikuroi* Nirenberg; sometimes mentioned in the literature as *F. moniliforme* Sheldon): producer of gibberellins, a group of hormone-like plant-growth stimulators; it was described originally as bakanae disease of rice in Japan. Gibberellins are now produced commercially as plant growth stimulators.

NOTE: Usually only the anamorphs are observed in culture.

***Gliocladium* Corda** **Fig. VIII.13d**
Anamorphic *Hypocreaceae*, Ascomycota

NUMBER OF TAXA: 12

TELEOMORPH: *Sphaerostilbella* (Henn.) Sacc. & D. Sacc.

REFERENCES: **Domsch et al. (2007)**, Kubicek & Harman (1998; industrial use), Rossman et al. (1999), **Seifert (1985)**, **Watanabe (2002)**

Examples: *G. penicillioides* Corda [Tel.: *Spaerostilbella aureonites* (Tul.) Seifert et al.]: conidial mass yellow to orange; parasitic on basidiomata of *Stereum* spp.; *G. viride* Matr.: conidia green in mass, in soil.

Note: Conidiophores are branched as in *Penicillium*, but the conidia are formed singly and not in chains, and form slimy droplets on the tip of the conidiophores. *Gliocladium* is used for enzyme production in industrial processes.

Glomerella Spauld. & v. Schrenk
Glomerellaceae, Ascomycota

Number of taxa: approx. 20

Anamorph: *Colletotrichum* Corda

References: von Arx (1970)

Examples: *G. cingulata* (Stonem.) Spauld. & v. Schrenk (Anam.: *Colletotrichum gloeosporioides* Penz.): apple rot. *G. tucumanensis* (Speg.) v. Arx & E. Müller [Anam.: *C. graminicola* (Ces.) Wilson)]: saprobe or parasite of Poaceae, in particular millet, corn and sugar cane.

Note: Most commonly only the anamorphs are found in culture.

Graphium Corda
Anamorphic *Microascaceae*, Ascomycota

Number of taxa: 17

Teleomorph: *Microascus* Zukal

References: Ellis (1971, 1976), Okada et al. (1998, 2000; neotypification, new definition of genus), Seifert & Okada (1993; definition of genus)

Note: Synnemata forming anamorphs of *Ophiostomataceae* with conidia aggregated in slimy heads at the apex of the synnemata were formerly assigned to *Graphium*. Okada et al. (2000) have shown, however, that the still valid *Graphium* species do not belong to the *Ophiostomataceae* but must be included in the *Microascaceae*.

Gymnoascus Baranetzki
***Gymnoascaceae*, Ascomycota**

NUMBER OF TAXA: 9

ANAMORPH: *Chrysosporium*- or *Malbranchea*-like

REFERENCES: **von Arx (1986)**, Domsch et al. (2007), Orr et al. (1963), Samson (1972)

EXAMPLE: *G. reesii* Baran.: i.a. soil fungus.

***Hainesia* Ell. & Sacc.**
Anamorphs *incertae sedis* and *Leotiomycetes*, Ascomycota

NUMBER OF TAXA: approx. 30

TELEOMORPH: *Discohainesia* Nannf.

REFERENCES: Sutton (1980), Sutton & Gibson (1977)

EXAMPLE: *H. lythri* (Desm.) Höhn.: leaf spot diseases and strawberry rot. The fungus is now known under the name *Discohainesia oenotherae* (Cooke & Ellis) Nannf.

***Hamigera* Stolk & Samson**

NOTE: No species of *Hamigera* are listed in *Index Fungorum*. They are currently included in *Emericella* [*Hamigera spinulosa* (Warcup) Arx; valid name: *Aspergillus spinulosus* Warcup], *Eupenicillium* (*Hamigera striata* (Raper & Fennell) Stolk & Samson; valid name: *Penicillium lineatum* Pitt) and *Talaromyces* [valid name: *Merimbla ingelheimense* (J. F. H. Beyma) Pitt]. Other species have not been revised so far.

***Hemicarpenteles* Sarbhoy & Elphick**
***Trichocomaceae*, Ascomycota**

NUMBER OF TAXA: 2

ANAMORPH: *Aspergillus* Sektion *Ornati*

REFERENCES: Samson (1979)

EXAMPLES: *H. ornatus* (Subram.) Arx., *H. paradoxus* A.K. Sarbhoy & Elphick.

Hormonema Lagerberg & Melin
Anamorphic _Dothioraceae_, Ascomycota

Number of taxa: 5

Teleomorph: *Dothiora* Fr.

Synanamorph: *Dothichiza* Lib.

References: De Hoog (1999), **Hermanides-Nijhof (1977)**

Example: *H. prunorum* (Dennis & Buhagiar) Herm.-Nijh.

Note: *H. dematioides* Lagerberg & Melin causes wood blue staining. The teleomorph is *Sydowia polyspora* (Bref. & Tavel) E. Müller. This fungus produces *Dothichiza pithyophila* (Corda) Petrak as a pycnidial synanamorph. *Hormonema* can be easily confused in culture with *Aureobasidium*, but its conidiogenesis is enteroblastic (basipetal formation). See also *Aureobasidium*.

Humicola Traaen **Fig. VIII.11h**
Anamorphic _Chaetomiaceae_, Ascomycota

Number of taxa: 17

References: De Bertoldi (1976), Domsch et al. (2007), Ellis (1971), Fassatiová (1967), Watanabe (2002)

Examples: *H. grisea* Traaen, *H. fuscoatra* Traaen: on plant remains and in soil.

Note: The fungus forms small, hyaline phialoconidia and large, thick-walled aleurospores.

Hypocrea Fr.
Hypocreaceae, Ascomycota

Number of taxa: ca. 100

Anamorph: *Trichoderma, Verticillium*

References: Chaverri & Samuels (2003), Doi (1972), Doi & Yamatoya (1989), Gams (2006), Jaklitsch et al. (2008), **Jaklitsch (2009)**, Rossman et al. (1999), Samuels et al. (1998)

EXAMPLE: *T. koningii* Lieckf., Samuels & W. Gams (anam.: *T. koningii* Oudem.): on rotten wood, cosmopolitan, mainly in temperate zones.

NOTES: Species of *Hypocrea* grow as saprobes on bark and wood, sometimes also as parasites of other fungi (e.g. Aphyllophorales).

Leptosphaerulina McAlpine
Incertae sedis, _Pleosporales_, Ascomycota

NUMBER OF TAXA: 13

REFERENCES: Aveskamp et al. (2010), Graham & Luttrell (1961)

EXAMPLE: *L. australis* McAlpine: saprobe on grasses.

Lichtheimia Vuillemin

NUMBER OF TAXA: 4

REFERENCES: de Hoog et al. (2000), St. Germain & Summerbell (2010) (both as *Absidia*). Web reference: http://www.zygomycetes.org/.

EXAMPLE: *L. corymbifera*: human pathogen; thermotolerant, rapidly growing between 37 and 42 °C (range 20–53 °C).

NOTE: *Lichtheimia* forms relatively small, apophysate sporangia with deliquescent walls; the sporangiophore has a single septum and stolons and rhizoids are produced; contrary to *Rhizopus*, the sporangiophore is not formed opposite the rhizoids. The zygospore has an ornamented wall and, contrary to *Absidia*, the suspensors are opposed and lack appendages. Many authors still include *Lichtheimia* in *Absidia*.

Mammaria Ces. ex Rabenh. **Fig. VIII.12e–f**
Anamorphic _Lasiosphaeriaceae_, Ascomycota

NUMBER OF TAXA: 1

TELEOMORPH: *Pseudocercophora* Subram. & Sekar.

REFERENCES: Domsch et al. (2007)

EXAMPLE: *Mammaria echinobotryoides* Ces.: on rotten plant remains; recently reported from cheese rinds produced in small Alpine cheese factories of Switzerland.

Melanconium **Link** **Fig. VIII.9c**

Anamorphic ***Melanconidaceae*****, Ascomycota**

NUMBER OF TAXA: approx. 50

TELEOMORPH: *Melanconis* Tul. & Tul.

REFERENCES: Sieber et al. (1991), Sutton (1980)

EXAMPLE: *Melanconium atrum* Lk.: on bark.

NOTE: A revision of the genus is badly needed.

Melanospora **Corda**

Ceratostomataceae**, Ascomycota**

NUMBER OF TAXA: approx. 25

REFERENCES: Cannon & Hawksworth (1982), Doguet (1955)

EXAMPLE: *M. chionea* (Fr.) Corda: on fallen pine needles.

NOTE: *Melanospora* species are common wood saprobes, but they are also isolated from dung or parasitize other fungi.

Microascus **Zukal**

Microascaceae**, Ascomycota**

NUMBER OF TAXA: 17

ANAMORPH: *Scopulariopsis*, *Wardomyces* Brooks & Hansford

REFERENCES: Abbott et al. (1998), **von Arx (1975a)**, Barron et al. (1961), Morton & Smith (1963)

EXAMPLES: *M. trigonosporus* Emmons & Dodge: in soil, on wood, cereal seeds, rice, dung. *Microascus brevicaulis* S. P. Abbott

[= *Scopulariopsis brevicaulis* (Sacc.) Bainier]: the most common species of this genus, with a cosmopolitan distribution and isolated from cereals (wheat), fruit (apples), soy, ground nuts, meat, cheese, milk and dairy products. Usually only the anamorphs are formed in culture.

NOTES: Some species (e.g. *S. brevicaulis*) are opportunistic human pathogens.

Microsporum Gruby — Fig. VIII.19b, c
Anamorphic *Arthrodermataceae*

NUMBER OF TAXA: 17

TELEOMORPH: *Arthroderma* Curr.

REFERENCES: **de Hoog et al. (2000)**, Gräser et al. (2008), **St. Germain & Summerbell (2010)**

EXAMPLE: *M. canis* (Bodin) Bodin: causal agent of ringworm in cats, dogs and monkeys; occasionally transmitted to humans, particularly children, and then causing tinea capitis and tinea corporis.

NOTES: *Microsporum* can be distinguished from *Epidermophyton* and *Trichophyton* by its echinulate, rough macroconidia. A similar dermatophyte, *Keratinomyces* Punsola & Guarro, has very thick-walled, smooth macroconidia and microconidia are always absent.

Monascus van Tieghem — Fig. VIII.6f–i
***Monascaceae*, Ascomycota**

NUMBER OF TAXA: 8

ANAMORPH: *Basipetospora*

REFERENCES: Domsch et al. (2007), Hawksworth & Pitt (1983), Samson et al. (2004)

EXAMPLE: *M. ruber* v. Tieghem (Anamorph. *Basipetospora rubra* Cole & Kendrick; synonym: *Monascus purpureus* Went.): reported from soil, boiled potatoes, rice, oat seeds, soy, millet, tobacco, stored grains. The fungus prefers substrates rich in starch and stored at temperatures higher than 25 °C.

NOTES: *M. ruber* is a strong producer of yellow to red pigments, depending on the nitrogen source. The fungus is used in China for the production of Angka, a red pigment derived from the fermentation of rice. It is also used to stain foodstuff in some Asian countries. It can stand short heat shocks (up to 96 °C). *M. ruber* produces also statins, blood cholesterol-lowering substances. Index Fungorum considers *M. ruber* and *M. purpureus* two distinct species, whereas MycoBank treats them as synonyms.

Monilinia **Pers.**
Sclerotiniaceae**, Ascomycota**

NUMBER OF TAXA: approx. 20

SYNONYM: *Sclerotinia* Fuckel

REFERENCES: Beyrde & Willetts (1977); CMI descriptions of pathogenic fungi: 616, 617, 618, 619; Fungi Canadenses: 38

EXAMPLES: *Monilinia fructigena* Honey: fruit rot of apples, pears, plums, peaches, and cherries. *Monilinia fructicola* (G. Winter) Honey: causal agent of brown rot of stone fruits, i.a. peaches, cherries, nectarines, and apricots. *Monilinia laxa* (Aderh. & Ruhland) Honey: brown rot of stone fruits.

NOTE: Species of *Monilinia* were formerly described in the mycological literature as *Monilia*.

Moniliella **Stolk & Dakin**
Incertae sedis**, Tremellomycetes, Basidiomycota**

NUMBER OF TAXA: 6

REFERENCES: Dakin & Stolk (1968), De Hoog (1979), Samson et al. (2004)

EXAMPLES: *M. acetoabutens* Stolk & Dakin: acidophilic, in vinegar containing preserved food, fruit juices, and syrups. *M. suaveolens* (Lindner) v. Arx: lipophilic and osmophilic, on margarine, in dairy products, causing black spots on cheese.

NOTES: The fungus forms arthroconidia as well as conidia in acropetal chains. See also *Aureobasidium*.

***Monographella* Petr.** **Fig. VIII.13e**
***Incertae sedis, Xylariales*, Ascomycota**

NUMBER OF TAXA: 7

REFERENCES: v. Arx (1984), Smiley et al. (2000), Uecker (1993)

EXAMPLE: *Monographella cucumerina* (Lindf.) Arx [= *Plectosphaerella cucumerina* (Lindf.) W. Gams, *P. tabacina* (J. F. H. Beyma) Palm et al. 1995]: frequent soil fungus, causing necrotic spots on potato stems; causal agent of the microdochium blight of cucumbers and occasionally causing wilt symptoms on carrots. *Monographella nivalis* (Schaffnit) E. Müll.: „pink snow mould" (sometimes called „Fusarium patch") on grasses. Both species are known in culture only as anamorphs.

***Mortierella* Coemans** **Fig. VIII.2h**
Mortierellaceae, Zygomycota

NUMBER OF TAXA: approx. 90

REFERENCES: Domsch et al. (2007), Gams (1976, 1977), Watanabe (2002), Zycha et al. (1969)

EXAMPLE: *M. bainieri* Cost.: often on rotting Basidiomycota, parasitically i.a. on button mushrooms, isolated also from forest soil.

***Mucor* Micheli : Fr.** **Fig. VIII.3a–c**
***Mucoraceae*, Zygomycota**

NUMBER OF TAXA: approx. 75

REFERENCES: **Domsch et al. (2007)**, Hermet et al. (2012; molecular taxonomy of cheese-colonising species), Samson et al. (2004), **Schipper (1978a)**, Zycha et al. (1969)

EXAMPLES: *M. fuscus* Bain.: on half-hard cheese from the Alps.
M. hiemalis Wehmer: the most common species, cosmopolitan and frequently isolated from soil, rapidly growing; tolerates high CO_2 concentrations (up to 100% CO_2). *M. plumbeus* Bon.: cosmopolitan, air-borne or in soil, on plant remains, hay, dung, spoilage of fruit-containing products and yoghurt. *M. racemosus* Fres.: cosmopolitan, very frequent in soil, on dung, in foodstuffs, cereal seeds and vegetables; the fungus forms heat-

resistant chlamydospores that can survive a heat treatment at 63 °C during 25 minutes.

NOTE: Some species are used for soy fermentation.

Mycogone Link **Fig. VIII.12d**

Anamorphic _Hypocreaceae_, Ascomycota

NUMBER OF TAXA: 5

TELEOMORPH: *Hypomyces* (Fr.) Tul.

REFERENCES: Barron (1968), Tubaki (1955)

EXAMPLES: *M. rosea* Link., *M. perniciosa* (valid name: teleomorph *Hypomyces perniciosus* Magnus): parasitically on Basidiomycota, i.a. on button mushrooms.

NOTE: *Mycogone* is characterised by the typical two-celled aleurospores; sometimes a *Verticillium* synanamorph is also present.

Nectria (Fr.) Fr.

Nectriaceae, Ascomycota

NUMBER OF TAXA: approx. 120, but additional species names to be found in the literature.

ANAMORPHS: (*Cylindrocarpon*), *Gyrostroma* Naumov, *Tubercularia* Tode, *Zythiostroma* Höhn.

REFERENCES: **Booth (1959)**, Hirooka et al. (2011), Rossman (1983, 1989), **Rossman et al. (1999)**, Samuels (1976, 1988, 1989), Samuels et al. (1991)

EXAMPLE: *N. cinnabarina* (Tode : Fr.) Fr.: on recently dead or dying twigs of several deciduous trees; very broad host range, common throughout the temperate zones.

NOTES: Species of *Nectria* are known plant pathogens and so far have never been observed as moulds. *Nectria sensu* Fr. is a genus complex that has been split into several genera. Rossman et al. (1999) have provided an excellent revision of the genera.

***Neocarpenteles* Udagawa & Uchiy.**
***Trichocomaceae*, Ascomycota**

NUMBER OF TAXA: 1

ANAMORPH: *Aspergillus* sect. *Clavati*

LITERATUR Udagawa & Uchiyama (2002), Varga et al. (2007)

EXAMPLE: *Neocarpenteles acanthosporum* (Udagawa & Takada) Udagawa & Uchiy [anamorph: *Aspergillus acanthosporus* Udagawa & Takada], in soil.

***Neopetromyces* Frisvad & Samson**
***Trichocomaceae*, Ascomycota**

NUMBER OF TAXA: 1

ANAMORPH: *Aspergillus* sect. *Circumdati*

REFERENCES: Frisvad & Samson (2000)

EXAMPLE: *Neopetromyces muricatus* (Udagawa, Uchiy. & Kamiya) Frisvad & Samson is an ochratoxin A producer.

***Neosartorya* Malloch et Cain** **Fig. VIII.7a–c**
***Trichocomaceae*, Ascomycota**

NUMBER OF TAXA: more than 20

ANAMORPH: *Aspergillus* sect. *Fumigati*

REFERENCES: Domsch et al. (2007), Kozakiewicz (1989), Samson et al. (1990), Samson et al. (2004), Samson et al. (2007)

EXAMPLE: *N. fischeri* (Wehmer) Malloch & Cain: thermotolerant, isolated from soil, rice, cotton, potatoes, ground nuts, leather and paper. Opportunistic pathogen of humans and animals. Ascospores germinate after a heat shock (60–65 °C during 30 min.). Good growth on media containing 15% NaCl.

NOTES: *Neosartorya* species are food spoilage fungi characterised by heart-resistant ascospores and able to colonise preserves.

Neurospora Shear et Dodge
Sordariaceae, Ascomycota

NUMBER OF TAXA: 13

ANAMORPH: *Chrysonilia*

REFERENCES: Moreau-Froment (1956), Perkins et al. (2000; genetics of the fungus), Samson et al. (2004)

EXAMPLES: *N. sitophila* Shear & Dodge: cosmopolitan, frequently on burnt plant remains. *N. crassa* Shear & Dodge: used in genetics.

NOTES: Usually only the anamorph *Chrysonilia* (= „*Monilia*") is observed in culture, the teleomorph is formed only in rare cases; the cultures are pink, exceptionally fast-growing (a 90 mm Petri dish is covered in less than 1 day!) and is thus a feared laboratory contaminant.

Paecilomyces Bainier — Fig. VIII.17a–c
Anamorphic *Trichocomaceae*, Ascomycota

NUMBER OF TAXA: >45; 10, all in sect. *Paecilomyces*, are important food spoilage fungi.

TELEOMORPH: *Byssochlamys*

REFERENCES: **Domsch et al. (2007), Samson (1974)**, Samson et al. (2004)

EXAMPLE: *P. variotii* Bainier: very frequently isolated from soil and air in warm regions; it may grow on a variety of substrates, including palm oil, cosmetics, pharmaceutical emulsions, ink, jute fabric, paper, leather, optic lenses, and drinks. The fungus forms heat-resistant aleurospores and can thus survive composting and pasteurisation. *P. variotii* is able to degrade plastics, PVC, and kerosene and is also resistant against several preservatives. It can also cause mycotoxicoses in humans and animals, being a producer of patulin and viriditoxin. In addition, it has been reported as an opportunistic pathogen in humans.

NOTES: *Paecilomyces* can be distinguished from *Penicillium* by the subulate phialides, the irregular penicilli as well as by cultural characters. *Paecilomyces* species are potentially all toxin producers, in particular of patulin.

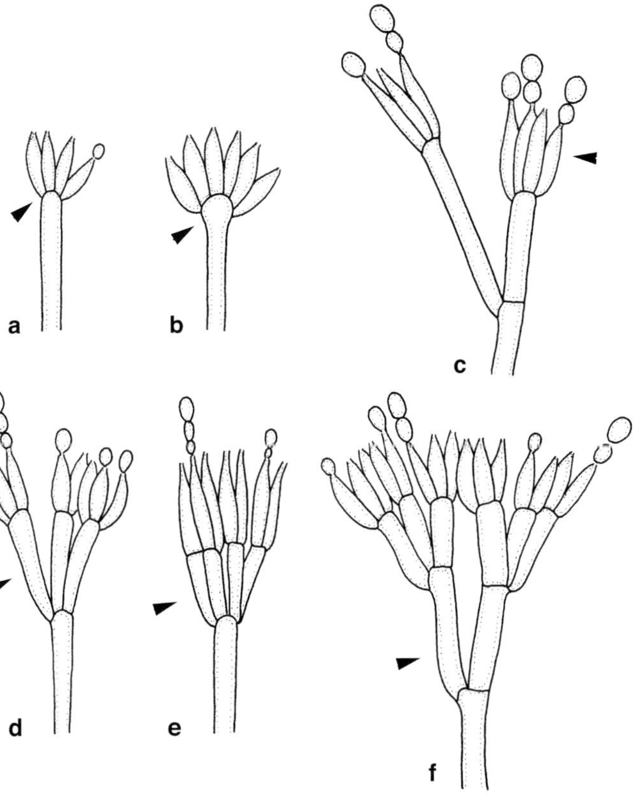

Fig. IX.2. Subgeneric classification of the genus* Penicillium *(Pitt, 2000)

a, b. Subgenus *Aspergilloides*, unbranched conidiophore: **a.** Not enlarged at the top (arrow), Sect. *Exicaulis*. **b.** Enlarged at the top (arrow), Sect. *Aspergilloides*. **c, d.** Subgenus *Furcatum*: **c.** conidiophore with limited, irregular branching under the apex, Sect. *Divaricatum* (arrow: phialide). **d.** Conidiophore branching only on one level, with phialides formed directly on metulae (sterile cells) that are usually longer than the phialides, Sect. *Furcatum*. **e.** Subgenus *Biverticillium*: conidiophore branching only on one level, with phialides formed directly on metulae that are usually as long as the phialides. **f.** Subgenus *Penicillium* (Terverticillata): conidiophore branching on two levels, with phialides formed on metulae that are disposed on rami (arrow).

***Penicillium* Link** **Fig. VIII.18, IX.2**

Anamorphic *Trichocomaceae*, Ascomycota

NUMBER OF TAXA: >100

TELEOMORPH: u.a. *Eupenicillium*, *Talaromyces*

REFERENCES: Dörge et al. (2000; image analysis applied to the identification), Domsch et al. (2007), Frisvad & Filtenborg (1989; chemotaxonomy), Mücke & Lemmen (2005; mycotoxins, health hazard), Peberdy (1987; biotechnology, secondary metabolites), **Pitt** (1974, 1979, **2000**), Pitt & Samson (1993; valid names and synonyms), Ramirez (1982), **Samson & Frisvad (2004**; terverticillate species**)**, **Samson & Houbraken (2011)**, Samson & Pitt (1985, 1990, 2000; molecular biology), **Samson et al.** (1976, **2004**), Singh et al. (1991), Stolk & Samson (1983)

EXAMPLES: *P. camemberti* Thom: used in the production of soft cheese. *P. chrysogenum* Thom: on foodstuff with low water activity, including nuts, dry fruit, fruit juices; growing on frescoes and optical lenses and air contaminant. *P. expansum* Link: causal agent of apple rotting and patulin producer. Some species are still used for penicillin production.

NOTES: Species identification of *Penicillium* isolates relies heavily on cultural characters. Strains must be grown on selective media at set conditions (CZY, MEA, 25 °C, 7 days). Differential characters are conidiophore branching (Fig. IX.2), structure, and dimensions; conidial shape and pigmentation, production of soluble pigments and exudates, and growth rate. TLC (detection of secondary metabolites) and PCR or sequencing (in particular for a rapid identification of the most common isolates) are also routinely used.

Currently running research projects aim at improving the molecular identification tools, in particular for species barcoding. MALDI-TOF is also increasingly used for the identification of the most common species.

Species of *Penicillium* are common soil inhabitants; they are cosmopolitan but reported preferentially from temperate zones. They may colonise a variety of substrates and their spores are air-borne. *Penicillium* species are excellent producers of a variety of useful secondary metabolites, producing (up to industrial scale) antibiotics such as penicillin and griseofulvin, but also toxins (e.g. citrinin, cyclopiazonic acid, patulin, roquefortin, viriditoxin). Several species are routinely used in biotechnology processes.

Petromyces Malloch & Cain
Trichocomaceae, Ascomycota

NUMBER OF TAXA: 1

ANAMORPH: *Aspergillus*

REFERENCES: Kozakiewicz (1989), Samson (1979)

EXAMPLE: *P. albertensis* J.P. Tewari

Pestalotiopsis Steyaert
Anamorphic _Amphisphaeriaceae_, Ascomycota

NUMBER OF TAXA: 22

TELEOMORPH: *Pestalosphaeria* M.E. Barr

REFERENCES: **Nag Raj (1993)**, Sutton (1980)

EXAMPLE: *P. funerea* (Desm.) Steyaert: cosmopolitan, on conifer needles; opportunistic pathogen of *Thuja*, causing dieback.

NOTES: Species of *Pestalotiopsis* are considered weakly pathogenic. The identification of species is extremely difficult, also because of the lack of reliable cultural studies. Most species were previously assigned to the genus *Pestalotia*, which was split into several genera by Sutton (1980) and Nag Raj (1993).

Dichotomous key to the _Pestalotiopsis_ genus complex

1. Conidiomata dome-shaped; conidia distoseptate .... *Pestalotia*
1.* Conidiomata are acervuli; conidia euseptate 2

2. Basal appendage, when present, formed exogenously ...... *Seimatosporium*
2.* Basal appendage, when present, formed endogenously 3

3. Conidia with only one apical appendage 4
3.* Conidia with more than one apical appendage 5

4. Conidia 3–4-septate .... *Monochaetia*
4.* Conidia 5-septate .... *Seiridium*

5. Conidia with 3 septa .... *Truncatella*

5.* Conidia with 4 septa .... *Pestalotiopsis*

***Phialophora* Medlar** **Fig. VIII.13f–i**

Anamorphic *Herpotrichiellaceae*, Ascomycota

Synonyms: i.a. *Margarinomyces* Laxa

Number of taxa: more than 20

Teleomorph: *Ascocoryne* Groves & Wilson, *Coniochaeta*, *Gaeumannomyces* von Arx & Olivier, *Mollisia* (Fr.) Karst., *Pyrenopeziza* Fuckel

References: Cole & Kendrick (1973), De Hoog (1999), **Domsch et al. (2007)**, Samson et al. (2004), **Schol-Schwarz (1970)**, **Watanabe (2002)**

Examples: *Ph. bubakii* (Laxa) Schol-Schwarz: isolated from butter and margarine, on which it forms small, black colonies. *Ph. cinerescens* (Wollenw.) Beyma causes fan-mould of carnation and can also infect strawberries. *Ph. mustea* Neergard is found in fruit juices, in particular apple juice; its heat-resistant aleurospores withstand pasteurisation.

Notes: Colonies are usually black or olive, creamy when young. Species belonging to the complex *P. hoffmanni* (Schol-Schwarz, 1970) are characterised by light-coloured, pinkish colonies and have been transferred to *Lecythophora* [e.g. *Lecythophora hoffmannii* (J.F.H. Beyma) W. Gams & McGinnis, Syn.: *Phialophora hoffmannii* (J.F.H. Beyma) Schol-Schwarz], anamorphic *Coniochaeta*. Not all species are characterised by a well-defined collarette, which should be the most distinctive character of the genus.

***Phlyctema* Desm.**

Anamorphic *Dermateaceae*, Ascomycota

Number of taxa: 4

Teleomorph: *Neofabraea* H. S. Jacks.

References: von Arx (1970), Sutton (1980)

EXAMPLE: *Phlyctema asparagi* Fautrey & Roum., *P. vagabunda* Desm. [valid name: teleomorph *Neofabraea alba* (E.J. Guthrie) Verkley]: saprobe and parasite on phanerogams, causing leaf spots; causal agent of apple rot.

Phoma Sacc. **Fig. VIII.9d–g**

Anamorphs _incertae sedis_ and _Pleosporales_, Ascomycota

NUMBER OF TAXA: approx. 2,000 (many names are probably synonyms)

TELEOMORPH: *Didymella* Sacc.

REFERENCES: Aveskamp et al. (2010), Boerema (1976, 1993), Boerema & De Gruyter (1998), Boerema & Dorenbosch (1973), Boerema et al. (1977, 1981, 1994, 1996, 1997), De Gruyter & Noordeloos (1992), De Gruyter et al. (1993, 1998), **Domsch et al. (2007)**, **Dorenbosch (1970)**, Monte et al. (1990; physiology, biochemistry), **Sutton (1980)**, White & Morgan-Jones (1983, 1986, 1987)

EXAMPLE: *P. glomerata* (Corada) Wollenw. & Hochapfel: isolated from soil, butter, rice grains, and fruits; causal agent of tomato rot.

NOTES: Extremely frequent and common genus, living as saprobes on different plants and plant remains, sometimes as parasites on plant organs and fruits. Some species have been isolated from soil and others have been described in the last few years as opportunistic human and animal pathogens. Species identification is very difficult.

Phomopsis (Sacc.) Bubàk **Fig. VIII.9h–j**

Anamorphic _Diaporthaceae_, Ascomycota

NUMBER OF TAXA: approx. 400

TELEOMORPH: *Diaporthe* Nitschke

REFERENCES: Sutton (1980), Uecker (1988), Wechtl (1990)

EXAMPLE: *Ph. occulta* (Sacc.) Traverso is a saprobe of conifers.

NOTES: Species of *Phomopsis* are often opportunistic plant pathogens, colonising deciduous trees, shrubs and grasses as well as conifers. No modern monograph is available and the species identification relies mainly on the host, from which the organisms have been isolated. *Phomopsis* often (but not always!) form two types of conidia, ellipsoid (alpha conidia) and filiform, straight or curved beta conidia, as well as intermediate forms of both.

***Phycomyces* Kunze : Fr.**
***Phycomycetaceae*, Zygomycota**

NUMBER OF TAXA: 3

REFERENCES: Benjamin & Hesseltine (1959), Cerdá-Olmedo & Lipson (1987; physiology, genetics, life cycle)

EXAMPLE: *Ph. blakesleeanus* Burgeff: very sensitive to light, gravity, wind, and chemicals, exhibiting a phototropic growth. It produces carotene and is widely used for physiological and genetic investigations.

NOTE: Species of *Phycomyces* can be easily isolated from fresh horse manure.

***Phyllosticta* Pers.**
Anamorphic *Botryosphaeriaceae*, Ascomycota

NUMBER OF TAXA: 46

TELEOMORPH: *Guignardia* Viala & Ravaz

REFERENCES: **van der Aa (1973)**, van der Aa & Vanev (2002), Bissett (1979, 1986a, 1986b), **Bissett & Palm (1989**; species on conifers), Leuchtmann et al. (1992; isoenzyme analysis), **L. Petrini et al. (1991**; species on conifers), Punithalingam (1974, 1981)

EXAMPLES: *P. ampelicida* (Engelman) van der Aa [valid name: teleomorph *Guignardia bidwellii* (Ellis) Viala & Ravaz]: grape black rot. *P. citricarpa* (Mc Alpine) van der Aa (valid name: teleomorph *Guignardia citricarpa* Kiely): black spots on *Citrus*.

NOTE: Several *Phyllosticta* species are the causal agents of black spots on many hosts.

Piptocephalis De Bary **Fig. VIII.1e, f**

Piptocephalidaceae, Zygomycota

Number of taxa: 24

References: Ellis & Ellis (1988), Zycha et al. (1969)

Example: *P. cylindrospora* Bain.: hyperparasite of other Zygomycota; isolated from rabbit dung.

Note: Most species are hyperparasites of Zygomycota or grow saprobically on dung.

Pithomyces Berk. et Broome **Fig. VIII.12g**

Anamorphs _incertae sedis_ and _Pleosporales_, Ascomycota

Number of taxa: 15

Teleomorph: *Leptosphaerulina*

References: Ellis (1960, 1971, 1976), Rao & De Hoog (1986)

Example: *P. chartarum* (Berk & Curt.) M.B. Ellis: on paper and dead plant remains, isolated from soil; known to cause a facial eczema in livestock (predominantly sheep).

Pseudogymnoascus Raillo

Pseudeurotiaceae, Ascomycota

Number of taxa: 8

Anamorph: *Geomyces* Traaen

References: Domsch et al. (2007), Orr (1979), Samson (1972)

Example: *P. alpinus* E. Müller: isolated from soil in extreme habitats (alpine regions).

Pyrenochaeta De Not.

Anamorphic _incertae sedis_ and _Pleosporales_, Ascomycota

Number of taxa: more than 120; most of them, however, probably belonging to other genera.

Teleomorph: *Herpotrichia* Fuckel

References: Sutton (1980)

Example: *Pyrenochaeta nobilis* var. *ilicis* (M. Wilson) R. Schneid.: on dead *Ilex aquifolium* leaves.

Ramichloridium Stahel ex de Hoog
Anamorphic _Mycosphaerellaceae_, Ascomycota

Number of taxa: 15

Teleomorph: *Capronia* Sacc.

References: Arzanlou et al. (2007), **De Hoog (1977)**, Domsch et al. (2007)

Example: *R. pini* de Hoog & Rahman: on dead *Pinus* branches.

Note: Usually plant pathogenic.

Rhinocladiella Nannfeldt
Anamorphic _Herpotrichiellaceae_, Ascomycota

Number of taxa: at least 7

Teleomorph: *Capronia* Sacc.

References: De Hoog (1977)

Example: *R. atrovirens* Nannf.

Note: Species of *Rhinocladiella* belong collectively to the black yeasts. Among them several taxa can be opportunistic human and animal pathogens, causing phaeohyphomycoses that are particularly refractory to treatment. See also *Aureobasidium*.

Rhizomucor Lucet & Costantin **Fig. VIII.3d, e**
Mucoraceae, Zygomycota

Number of taxa: 4

References: Domsch et al. (2007), De Hoog et al. (2000), Schipper (1978b)

Example: *R. pusillus* (Lindt) Schipper (Syn.: *Mucor pusillus* Lindt): worldwide distribution, isolated from soil; it withstands heat treatment at 68 °C during 45 min. Opportunistic human and animal pathogen.

NOTE: *Rhizomucor* species can be distinguished from *Mucor* by the presence of stolons and rhizoids; in addition, they are thermophilic.

***Rhizopus* Ehrenb.** **Fig. VIII.4**
***Mucoraceae*, Zygomycota**

NUMBER OF TAXA: 10

REFERENCES: **Dabinett & Wellman (1973)**, **De Hoog et al. (2000)**, Liu et al. (2007; molecular phylogeny), Samson et al. (2004), Schipper (1984), Schipper & Stalpers (1984), **Zheng et al. (2007, monograph)**

EXAMPLES: *R. oligosporus* Saito: thermophilic, used for the preparation of Tempe (Asian foodstuff). *R. oryzae* Went & Prinsen Geerlings: thermophilic, isolated mainly in tropical and subtropical regions from soil, cereals, ground nuts, polluted water, vegetables and rotting fruits. *R. stolonifer* (Ehrenb.: Fr.) Vuill.: in soil and from cereals, vegetables, fruits, and nuts.

NOTES: *Rhizopus* can be distinguished from *Mucor* by its stolons and rhizoids, as well as by the typical apophysis. Some species are opportunistic human and animal pathogens.

***Scedosporium* Sacc. ex Castell. et Chalm.**
Anamorphic *Microascaceae*, Ascomycota

NUMBER OF TAXA: 4

TELEOMORPH: *Pseudallescheria* Negr. et I. Fisch, *Petriella* Curzi

REFERENCES: **De Hoog et al. (2000)**, Gilgado et al. (2006), **Guarro et al. (2006)**, **St-Germain & Summerbell (2010)**

EXAMPLES: *S. prolificans* (Hennebert et Desai) Guehó et de Hoog: isolated from warm, indoor soil; it may enter the human body through lesions caused by thorns and splinters and is frequently isolated from cutaneous and subcutaneous lesions. In immunocompromised patients it may cause fatal infections.

NOTE: *Scedosporium* species are known human pathogens; they can be distinguished from *Scopulariopsis* by the slimy conidial heads and the lack of conidial chains.

Scopulariopsis Bainier **Fig. VIII.17d–g**
Anamorphic _Microascaceae_, Ascomycota

Number of taxa:	approx. 20
Teleomorph:	*Microascus*
References:	**Morton & Smith (1963)**, Domsch et al. (2007), Abbott et al. (1998; teleomorph), **De Hoog et al. (2000)**, Samson et al. (2004)
Examples:	*S. brevicaulis* (Sacc.) Bainier (cf. *Microascus*). *S. candida* (Guéguen) Vuill.: isolated i.a. from soil, rice, paper, cheese, nails and skin. *S. flava* (Sopp) F.J. Morton & G. Sm.: i.a. on the rind of hard cheese. *S. fusca* Zach: on cheese rind, in chicken eggs, on cacao leaves; the species can penetrate hair (production of keratinases).
Note:	Several *Scopulariopsis* species are known human pathogens, causing i.a. onychomycoses.

Sepedonium Link
Anamorphic _Hypocreaceae_, Ascomycota

Number of taxa:	approx. 10
Teleomorph:	*Hypomyces* (Fr.) Tul.
References:	Barron (1968), Domsch et al. (2007)
Example:	*S. chrysospermum* (Bull.) Link (valid name: teleomorph *Hypomyces chrysospermus* Tul. & C. Tul).: parasitic on *Boletales* and *Agaricales* (Basidiomycota), sometimes called „bolete eater" or „yellow mould".

Sordaria Ces. et de Not. **Fig. VIII.7g–j**
Sordariaceae, Ascomycota

Number of taxa:	more than 20
References:	Domsch et al. (2007), Ellis & Ellis (1988), **Lundqvist (1972)**
Example:	*S. fimicola* (Rob.) Ces. & De Not.: mostly found world-wide

in the temperate zones, coprophilic, sometimes isolated from soil. Homothallic and with short reproductive cycles (approx. 10 days), therefore used for genetic investigations. Active phototropic ascospore release.

NOTE: Most *Sordaria* species are coprophilic.

Spilocaea Fries
Anamorphic *Venturiaceae*, Ascomycota

NUMBER OF TAXA: 2 (probably additional species).

TELEOMORPH: *Venturia* Sacc.

REFERENCES: **Ellis** (1971, **1976**), Sivanesan (1977, 1984)

EXAMPLE: *Sp. pomi* Fr. [valid name: teleomorph *Venturia inaequalis* (Cooke) G. Winter]: Apple scab disease.

NOTE: *Spilocea* species are usually described in monographs and investigations on *Venturia*.

Stachybotrys Corda — Fig. VIII.13j–k
Anamorphs *incertae sedis* and *Hypocreales*, Ascomycota

NUMBER OF TAXA: more than 20 (additional species are likely).

REFERENCES: **Domsch et al. (2007), Ellis (1971)**, Gravesen et al. (1994, discussion of spoilage by selected species), Samson et al. (2004)

EXAMPLE: *S. chartarum* (Ehrenb.) Hughes (Syn.: *S. atra* Corda): cosmopolitan, isolated from dead plant material, hay, plaster, paper, clothing, wheat and oat seeds, and ivy. The species is strongly cellulolytic and grows at high water pressure (a_w>0.94), thus being commonly found in water damaged or very humid buildings.

NOTES: Species of *Stachybotrys* may produce several mycotoxins with strongly immune supressing properties (i.a. macrocyclic trichothecenes and antranones), as well as large amounts of spirocyclic compounds with fibrinolytic activity that are used in the pharmaceutical industry. The toxins produced by *Stachybotrys* are quite stable and can remain active, e.g. in the hay, over long periods of time, causing toxicoses in the animals ingesting the contaminated feed. The presence of *Stachybotrys* in a building is indicative of high humidity.

Stagonospora (Sacc.) Sacc.
Anamorphic *Phaeosphaeriaceae*, Ascomycota

NUMBER OF TAXA: more than 100

TELEOMORPH: *Phaeosphaeria* I. Miyake

REFERENCES: Castellani & Germano (1977), Hsieh (1979), Smiley (2000, pathogenic species on turfgrass), Sutton (1980)

EXAMPLES: *St. nodorum* (Berk.) Castellani & Germano [valid name: teleomorph *Phaeosphaeria nodorum* (E. Müller) Hedjaroude]: a major pathogen of wheat, on which it causes the so-called glume blotch.

NOTES: Species of *Stagonospora* are frequent on grasses and are important pathogens of cereals.

Stemphylium Wallr.
Anamorphic *Pleosporaceae*, Ascomycota

NUMBER OF TAXA: 14 (more not fully characterised)

TELEOMORPH: *Pleospora*

REFERENCES: Ellis (1971, 1976), Simmons (1967, 1969, 1985)

EXAMPLE: *S. botryosum* Wallr. (valid name: teleomorph *Pleospora tarda* E. Simmons): cosmopolitan, growing saprobically or weakly pathogenically on several plant species.

Syncephalastrum Schröter
***Syncephalastraceae*, Zygomycota**

NUMBER OF TAXA: 5

REFERENCES: Domsch et al. (2007), Samson et al. (2004), Zycha et al. (1969)

EXAMPLE: *S. racemosum* Cohn ex J. Schröt.: tropical and subtropical species, isolated from soil, dung, cereals, and spices; it may survive but will not grow under anaerobic conditions.

Talaromyces C. R. Benj.
Trichocomaceae, Ascomycota

NUMBER OF TAXA: approx. 30

ANAMORPH: *Merimbla* Pitt, *Penicillium*

REFERENCES: **Domsch et al. (2007)**, Frisvad et al. (1990; chemotaxonomy), **Pitt (1979, 2000)**, Samson et al. (2004, **2011**), Stolk (1965), Stolk & Samson (1972)

EXAMPLE: *T. flavus* (Klöcker) Stolk & Samson var. *flavus*: world-wide distribution, isolated from soil, organic material and tropical fruit juices; heat-resistant, the ascospores surviving in the soil up to 30 min. at 70 °C, in apple juice 40 min. at 80 °C, whereas the cleistothecia may survive 30 min. at 100 °C. Oatmeal agar is a good substrate for isolation and cultivation. The fungus produces some mycotoxins, some with antibiotic activity.

NOTE: One *Hamigera* species (*Hamigera avellanea* Stolk & Samson, Anamorpic *Talaromyces*, valid name *Merimbla ingelheimense*) has been transferred to *Talaromyces*.

Thamnidium Link **Fig. VIII.1a–d**
Mucoraceae, Zygomycota

NUMBER OF TAXA: 1

REFERENCES: Domsch et al. (2007), Hesseltine & Anderson (1956), Zycha et al. (1969)

EXAMPLE: *T. elegans* Link: psychrophilic and coprophilic, common in cold regions and isolated from soil, cereal seeds, air or dust; occasionally found in foodstuff, e.g. stored meat or goat cheese. Sporulation is induced by the light and the fungus is able to grow at 1–2 °C.

Thermoascus Miehe
Thermoascaceae, Ascomycota

NUMBER OF TAXA: 6

ANAMORPH: *Paecilomyces*, *Polypaecilum* G. Sm.

REFERENCES: Domsch et al. (2007), **Samson (1974)**, Stolk (1965)

Example: *Th. crustaceus* (Apinis & Chester) Stolk: thermophilic soil fungus, occasionally isolated also from pasteurised drinks.

Note: All known *Thermoascus* species are thermophilic.

Thermomyces Tsiklinsky
Anamorphs *incertae sedis* and *Eurotiales*, Ascomycota

Number of taxa: 4

References: Apinis & Eggins (1966), Cooney & Emerson (1965), Domsch et al. (2007), Ellis (1971), Pugh et al. (1964)

Example: *T. lanuginosus* Tsiklinsky: thermophilic soil fungus, i.a. growing in compost during pasteurisation. Minimum growth temperature is 30 °C, with an optimum at 40–55 °C. It produces enzymes (including esterases) that may degrade plastics.

Thielavia Zopf
***Chaetomiaceae*, Ascomycota**

Number of taxa: 25

Anamorph: *Acremonium*-like, *Myceliophthora* Costantin

References: **Domsch et al. (2007)**, Malloch & Cain (1973), von Arx (1975b)

Example: *T. heterothallica* Klopotek: thermophilic fungus, isolated from compost. *Th. terricola* (Gilman & Abbott) Emmons: distribution limited to warm climates, coloniser of soil, dung, seeds, rotting fruits; able to degrade chitin and cellulose.

Thielaviopsis Went. **Fig. VIII.15h–i**
Anamorphic *Ceratocystidaceae*, Ascomycota

Number of taxa: 1

Teleomorph: *Ceratocystis* Ellis & Halst.

References: Ellis (1971)

Example: *Th. basicola* (Berk. & Br.) Ferraris (= Syn.: *Chalara elegans* Nag Raj & Kenrick): black root rot of a wide range of plants, including carrots, chicory, fenugreek, and tobacco.

Tolypocladium W. Gams
Anamorphic _Ophiocordycipitaceae_, Ascomycota

NUMBER OF TAXA: 10

TELEOMORPH: *Elaphocordyceps* G.H. Sung & Spatafora

REFERENCES: Gams (1971b), Sung et al. (2007; phylogeny)

EXAMPLE: *T. cylindrosporum* W. Gams: soil fungus.

NOTE: *Elaphocordyceps subsessilis* (Petch) G. H. Sung, J. M. Sung & Spatafora is the valid name of *T. inflatum* W. Gams, the producer of the immune suppressant cyclosporin.

Torula Pers.
Anamorphic _incertae sedis_, Ascomycota

NUMBER OF TAXA: 16

TELEOMORPH: Pezizomycotina

REFERENCES: Domsch et al. (2007), Ellis (1971), **Rao & De Hoog (1975)**

EXAMPLE: *T. ligniperda* (Willk.) Sacc. grows as dark colonies on dead wood; it causes wood staining in timber of deciduous trees.

NOTE: Species of *Torula* live mostly saprobically on dead plant material; some have been isolated from dung and soil.

Trichoderma Pers. **Fig. VIII.15a**
Anamorphic _Hypocreaceae_, Ascomycota

NUMBER OF TAXA: approx. 130

TELEOMORPH: *Hypocrea* Fr.

REFERENCES: **Bissett (1984, 1991a, b, c)**, Chaverri et al. (2003), Degenkolb et al. (2008), De Respinis et al. (2009), Domsch et al. (2007), Druzhinina et al. (2006, 2008), Gams (2006), Jaklitsch et al. (2006, 2008), **Jaklitsch (2009)**, Kindermann et al. (1998), Klein & Eveleigh (1998), Komon-Zelazowska et al. (2007), Kraus et al. (2004), Kubicek & Harman (1998; industrial applications), Kubicek et al. (2008), Kuhls et al. (1997), Lu et al. (2004), **Rifai (1969)**, Samson et al. (2004), Samuels (2006),

Samuels & Ismaiel (2009), Samuels et al. (1998, 1999, 2002, 2006a, 2006b)

EXAMPLES: *T. aggressivum* Samuels et al. hinders mycelial growth of *Agaricus bisporus* in the cultivations. *T. harzianum* Rifai has been isolated from soil, paper, clothing, cereal seeds, nuts; strong cellulase and amylase producer. *T. viride* Pers. is characterised by ornamented (warted) conidia; wood coloniser, isolated also from soil, stored grain, nuts, tomatoes, sweet potatoes, citrus fruits.

NOTES: *Trichoderma* species as described by Rifai (1969) are taxonomic complexes that have now been split in several species, some of them cryptic (see the more recent taxonomic literature references above), using mainly molecular biology and proteomic techniques. Some *Trichoderma* species are used in biological control programs and are widely employed in industrial processes as cellulose producers. Opportunistic human pathogens have been recently reported.

Trichophyton Malmsten — Fig. VIII.19d-h

Anamorphic Arthrodermataceae

NUMBER OF TAXA: more than 100 (many possibly synonyms)

TELEOMORPH: *Arthroderma* Curr.

REFERENCES: **de Hoog et al. (2000)**, Gräser et al. (2008), **St. Germain & Summerbell (2010)**

EXAMPLE: *T. mentagrophytes* complex (Teleomorphs: *A. vanbreuseghemii* Takashio, *A. benhamiae* Ajello et S. L. Cheng) is a cosmopolitan species complex; anthropophilic isolates may cause chronic infection of the feet ("athlete's foot"), nails and groin. Zoophilic isolates may be the agents of lesions of the scalp, skin, nail, and beard regions.

NOTES: *Trichophyton* differs from *Microsporum* by its smooth, mostly thin-walled macroconidia and from *Epidermophyton* by the more regular shape of the macroconidia; microconidia are always absent in *Epidermophyton*, but they are not always formed in *Trichophyton*, making this a conditional distinctive character only.

***Trichothecium* Link** **Fig. VIII.15b–c**

Anamorphic *Hypocreales*, Ascomycota

NUMBER OF TAXA: 5

REFERENCES: Barron (1968), Domsch et al. (2007), Samson et al. (2004), **Summerbell et al. (2011)**

EXAMPLE: *T. roseum* (Pers.) Link: world-wide distribution, on dead plant material, on basidiomata of Basidiomycota, isolated from soil, cereal seeds, foodstuff, meal; may cause apple fruit rot. Spores are air-borne, fast growth in culture.

***Ulocladium* Preuss** **Fig. VIII.12h**

Anamorphic *Pleosporaceae*, Ascomycota

NUMBER OF TAXA: 14

REFERENCES: Ellis (1971, 1976), Samson et al. (2004), Simmons (1967)

EXAMPLE: *U. chartarum* (Preuss) Simmons: isolated from soil, dung, dead plant material, wood, air, dust, paper, clothing; it may grow on humid tapestry and paint and is known to cause upper respiratory tract allergies.

NOTE: *Ulocladium* may be distinguished from *Alternaria* by its conidia lacking a beak; the conidium tip elongates and becomes a conidiogenous cell; in most species no chains are formed.

***Verticillium* Nees**

Anamorphic *Plectosphaerellaceae*, Ascomycota

NUMBER OF TAXA: approx. 40

REFERENCES: **Domsch et al. (2007), Gams (1971a)**, Gams & Zare (2002), Rossman et al. (1999), Tjamos et al. (2000, phytopathology)

EXAMPLE: *V. albo-atrum* Reinke & Berthold is common in the northern temperate zone, in roots and is the causal agent of the Verticillium wilt in several host plants (e.g. potatoes, tomatoes). *V. dahliae* Kleb. has been reported from the temperate and subtropical regions: it also inhabits roots and may cause wilt. It can be distinguished from *V. albo-atrum* by the formation of microsclerotia.

Wallemia Johan-Olsen — Fig. VIII.10d–g

Anamorphic *Wallemiaceae*, Basidiomycota

Number of taxa: 3

References: Ellis (1971), Domsch et al. (2007), Samson et al. (2004), Zalar et al. (2005)

Example: *W. sebi* (Fr.) v. Arx is cosmopolitan, osmophilic and xerophilic: it has been isolated from soil and air. It colonises dry foodstuff, preferentially preserves, dates, bread, pies, nuts, salted fish, salted beans, sweetened condensed milk, and dry fruits.

Xeromyces Fraser

***Monascaceae*, Ascomycota**

Number of taxa: 1

Anamorph: *Fraseriella bispora* Cif. & Corte

References: Hawksworth & Pitt (1983), Pitt & Hocking (1983), Samson et al. (2004)

Example: *X. bisporus* L. R. Fraser is a particularly xerophilic, rather uncommon fungus, needing a minimum a_w of 0.61, thus growing preferentially on media with a high sucrose content; it has been isolated from dry foodstuff, dates, fruit pies, puddings. It tolerates very low oxygen pressures; the ascospores are heat-resistant and can survive temperatures of 80 °C.

Zygorhynchus Vuill. — Fig. VIII.3f–h

***Mucoraceae*, Zygomycota**

Number of taxa: 9

References: Hesseltine et al. (1959), Domsch et al. (2007)

Example: *Z. moelleri* Vuill.: world-wide distribution, found in soils from lowland to the alpine region; it grows also at high CO_2 concentrations.

Note: *Zygorhynchus* is characterised by zygosporangia subtended by two asymmetric zygophores.

X. Glossary

Acervulus	Flat, open fruit body (conidioma) of anamorphic Ascomycota or Basidiomycota (Fig. VIII.9a, c, j)
Acicular	Needle-shaped
Acroauxic	Growth of the conidiophore (in length) is restricted to its apical part (antonym: basauxic)
Acrogenous	Developing at the apical part (antonym: basigenous)
Acrokont	(Of a flagellum) inserted on the apical part of the zoospore
Acropetal chain	A chain of conidia in which the youngest conidium is formed at the tip of the chain. Acropetal chains are often branched.
Adelophialides	Needle-shaped phialides with the appearance of a denticle
Aleurospore (Aleuroconidium)	Thick-walled, vegetative resting spore formed at the hyphal tips. Morphologically they are almost indistinguishable from the chlamydospores (Fig. VIII.5d, VIII.13f).
Allantoid	Sausage-shaped
Amerospor	One-celled (aseptate) spore
Amyloid	(Of a structure, for example the ascal plug in Ascomycota) staining blue in iodine or Melzer's reagent
Anamorph	The asexual reproductive form of a fungus (see also teleomorph)
Anellide, anellidic	Conidiogenous cells growing percurrently and elongating, thus forming ring-like structures at the opening, sometimes conspicuous, more often, however, difficult to observe. Anelloconidia are usually easily recognised by their broad, flattened base. They often form basipetal chains (Fig. VIII.15f, g; VIII.17d, e g).
Antheridium (pl. antheridia)	Male gametangium
Anthropophilic	(Of dermatophytes) colonising preferentially humans
Apophysis	(Zygomycota) a funnel-shaped enlargement of the sporangiophore at the basis of the columella (Fig. VIII.4e)

Apothecium (pl. apothecia)	(Ascomycota) Disk-shaped, saucer-shaped or cup-shaped ascoma of some Ascomycota. Apothecia can be sessile or stipitate (with a stipe).
Appendiculate	With an appendage
Appressorium (pl. appressoria)	Special pressing organs produced on the germ tubes of some fungi, from which a needle-like peg is formed. This punctures the epidermis of the host, thus allowing hyphal penetration and the development of a mycelium inside the host. I.a. typical of *Colletotrichum* species.
Arthric	Kind of conidial ontogeny, by which the whole hypha breaks down into several conidia (Fig. VIII.10).
Ascogenous hyphae	Dikaryotic hyphae giving raise to the asci
Ascogonium (pl. ascogonia)	The female reproductive organ of Ascomycota
Ascoma (pl. ascomata)	The fruit body of Ascomycota, comprising different types (apothecium, cleistothecium, gymnothecium, perithecium, pseudothecium)
Ascospore	The meiotic spore of Ascomycota, formed in the asci
Ascus (pl. asci)	Meiosporangium (a sporangium in which meiosis occurs) of the Ascomycota (Fig. 7a, j)
Ascus tip (apical tip)	The apical part of the ascus (Fig. VIII.7j). In the unitunicate genera, ascospores are ejected actively through it (Fig. VII.2.1g). The ascus tip can be variously shaped and is sometimes provided with a distinct apical plug.
Awl-shaped	With an abrupt narrowing; typical of some phialides (Fig. VIII.5b–c, VIII.17a–b)
Azygospore	A zygospore developing parthenogenetically (reproduction by the development of an unfertilized spore)
Bacilliform	Rod-shaped
Basal stroma	A stroma on which the ascomata or conidiomata are seated.
Basauxic	Growth of the conidiophore (in length) is restricted to its basal part (antonym: acroauxic)
Basidiospore	Meiospore (sexual spore) formed on basidia (Fig. VIII.8b)

Basidium (pl. basidia)	Meiosporangium of the Basidiomycota. Basidia form exogenously basidiospores (Fig. VIII.8c).
Basipetal chain	A chain of conidia with the oldest conidium at the apex and the youngest at the base (antonym: acropetal chain). Basipetal chains are never branched.
Biconical	With the shape of two cones placed together apex to apex
Bitunicate	Bitunicate asci have a wall that appears two-layered in light microscopy (electron microscopy reveals that the wall is actually multi-layered). The two layers have different properties with regards to their elasticity (Fig. VII.2.1h; VIII.8a).
Biverticillate (in the genus *Penicillium*)	Two (rarely three) levels of branching directly below the phialides: phialides are seating directly on sterile cells called metulae, which in turn are disposed on the rami of the conidiophores (Fig. VIII.18b–c, IX.2c–e).
Blastic	Conidia are produced by budding from a conidiogenous cell (Fig. VIII.11–18).
Blastoconidium	Conidium formed blastically
Catenate, catenulate	In chains
Cheiroid	Hand-shaped
Chitinoid	A substance with the properties of chitin that cannot be stained with iodine or Melzer's reagent, staining however with cotton blue or ink.
Chlamydospore	Thick-walled, often melanised hyphal cells with a very variable shape (Fig. VIII.9g; VIII.14b). They are resting, long-living and resistant organs that are able to store nutrients and protect the fungus' genetic material from ecological and chemical damage.
Chondroid	Hyphae with thick, swelling walls, as seen in *Acremonium* Sect. *Gliomastix*
Chromophil	Structures (phialides, chlamydospores, conidia) with gelatinous, warty, strongly staining incrustations (*Acremonium*; Fig. III.13a)
Cicatrized	With scars; of a region on a conidiogenous cell with conspicuous scars (site of conidial formation)
Ciliate	With cilia (thread-like or bristle-like appendages)

Cirrus	A mucus-bound, ribbon-like mass of spores exuded from fungal fruit body
Clamp connection	A hyphal outgrowth in many Basidiomycota, formed at cell division over the cross wall between two cells and providing a connection through which one of the involved nuclei may pass to the other; called also buckle (s. Fig. II.3.1b).
Clavate	Club-shaped
Cleistothecium (pl. cleistothecia)	More or less spherical, closed fruit body of the Ascomycota. Their wall (peridium) will rupture when asci are ripe, thus allowing their dispersal. The cleistothecial wall may be composed of more or less tightly interwoven hyphae (they are then called gymnothecia) or have a cellular structure (Fig. VII.2.1a; VIII.5g; VIII.6f).
Clypeus	A shield-shaped structure covering one or more ascomata
Coenocytic	Hyphae lacking cross walls, typical of Zygomycota and some fungus-like Protista (Oomycota). Some Zygomycota, however, can form cross walls when old; the zygomycetous hyphae, however, are rather broad (4–20 µm), whereas other fungi seldom have such broad hyphae.
Collarette	A columnar or flared structure at the apex of a phialide, consisting of the outer wall layer
Columella	A sterile (non-reproductive) structure that extends into and supports the sporangium in the Mucorales (Fig. VIII.13e, g–i). It arises as a swelling of the distal part of the sporangiophore. Its function has not been conclusively elucidated but it is generally considered to be of structural support.
Conidiogenesis, conidial ontogeny	The formation of conidia
Conidiogenous cell	The cell from which conidia originate
Conidioma (pl. conidiomata)	Fruit body of anamorphic fungi (Coelomycetes) (Fig. VIII.8d, f, h; VIII.9a, c d, j)
Conidiophore	The structure bearing conidiogenous cells. Conidiophores may consist of only one cell subtending the conidiogenous cell or form a complex, differentiated structure.
Conidium	Sessile spore formed asexually (by mitosis: mitospore)

Coremium (pl. coremia)	See Synnema
Crozier	The recurved tip of ascogenous cells. After simultaneous division of the two nuclei in this cell the crozier will develop into a uninucleate apical cell (the recurved part of the hyphal tip) and a subtending cell immediately behind it. In the crozier karyogamy and meiosis occurs and from it asci and ascospores will develop.
Cuneiform	Wedge-shaped (see also spatulate)
Cupulate	Shaped like a cupule (inversed cup-shaped structure)
Determinated	Conidiogenous cell not growing any further after conidium formation (antonym: undeterminated)
Dichotomous	A system of branching in which the main axis forks repeatedly into two branches (Fig. VIII.1b, c, f)
Dictyosporous	With longitudinal and transversal septa (Fig. VIII.11a; VIII.9a, b)
Didymosporous	Two-celled spore
Dikaryon	Fungal hypha containing cells with two haploid nuclei (i.a. Basidiomycota)
Dikaryotic	A cell with two nuclei
Dimorphism, dimorphic	Dimorphic fungi grow either yeast-like or form hyphae. Switching from the budding yeast to the filamentous form is mainly triggered by ecological factors. Several dimorphic fungi are facultative human or animal pathogens: the parasitic form is mainly yeast-like, while the saprobic phase is filamentous.
Diploid	A nucleus containing two sets of chromosomes (usually, one set from each parental cell)
Discomycetes	Ascomycota with unitunicate asci contained in apothecia
Discrete (conidiogenous cell)	The conidiogenous cell can be distinguished from other cells of the conidiophore (antonym: integrated).
Disjunctor cell	An empty cell, with no nucleus, situated between two arthrocondia and undergoing lysis to release them

Distoseptate	A thin septum penetrated by cytoplasmic junctions, possibly formed by invagination from the lateral wall (e.g. *Drechslera*) (antonym: euseptate)
Doliiform	Barrel-shaped
Doliporus	A type of septum found in many Basidiomycota, with a pore surrounded by a thickened rim and caps on both sides (s. Fig. II.3.1d)
Echinulate	Spiny, spiky
Ectostroma	The outer part of a stroma, often with a different texture than the inner part (entostroma); usually well differentiated in wood-inhabiting fungi
Elliptic, ellipsoidal, ellipsoid	With the shape of an ellipse in section
Endogenous	Produced within a cell or a hypha
Endophytes	Fungi living within the tissues of their hosts
Endostroma	The inner part of a stroma, in which the perithecia are formed
Enteroblastic (conidiogenesis)	A conidiogenesis characterised by the conidial cell wall being formed by the inner layers of the conidiogenous cell walls only. The conidiogenous cell can be likened then to a bottle, from which conidia are released in sequential manner (Fig. VIII.13–18).
Epiphytes	Fungi living on the surface of their hosts
Epithecium	The surface layer of an apothecium, usually covering the hymenium
Eukaryotic	Cells containing a nucleus enclosed in a nucleic membrane
Euseptate	With well-defined septa; a true septum formed from a septal plate (antonym: distoseptate)
Eutunicate	An ascus with an active spore discharge, with a firm membrane and usually composed by a stipe and an ascospore containing part (Fig. VIII.7j; antonym: prototunicate)
Excipulum	External lateral layer surrounding an apothecium
Exogenous	Formed outside of an organ or cell
Falcate	Sickle-shaped

Fasciculate	Bundled
Filiform	Thread-like (Fig. VIII.9i)
Foot cell	The distinct basal cell of multicellular macroconidia in the genus *Fusarium* (Fig. VIII.14c). The shape of the foot cell in *Fusarium* macroconidia is an important species delimiter within the genus.
Fusoid	Spindle-shaped
Geophilic	(Of dermatophytes) growing usually in the soil and infecting occasionally humans and animals
Germ pore	A small pore in the outer wall of a fungal spore through which the germ tube exits upon germination. A germ pore can usually be distinguished as a light spot on otherwise brown spores (Fig. VIII.7f, h).
Germ slit, germ fissure	A longitudinal slit in the outer wall of a fungal spore through which the germ tube exits upon germination. A germ slit can usually be distinguished as a light fissure on otherwise brown spores (Fig. VIII.12e–f).
Glabrous	Without setae or hairs, smooth
Gymnothecium	Cleistothecium in which the cleistothecial wall may be composed of more or less tightly interwoven hyphae; typical of Ascomycota with prototunicate asci (e.g. *Gymnoascus* and *Pseudogymnoascus*).
Haploid	Cell with only one set of chromosomes
Helicosporous	Coiled, often multicellular spore
Heterokaryotic	Of cells with two or more genetically different nuclei sharing a common cytoplasm (antonym: homokaryotic)
Hilum	A more or less melanised scar on a spore, indicating the point of attachment to the conidiogenous cell
Holoblastic	The conidia burst at the tip of the conidiogenous cell and all cell wall layers are involved in conidium development (Fig. VIII.11, 12; antonym: enteroblastic).
Holomorph	"The whole fungus", composed of anamorph and teleomorph

Homokaryotic	Of cells with two or more genetically identical nuclei sharing a common cytoplasm
Hyaline	Colourless, almost transparent
Hymenium	The fertile layer in a fruiting body, consisting of asci or basidia
Hyphopodium (pl. hyphopodia)	A short, lobed (usually one-celled, sometimes two-celled) outgrowth from a hypha of epiphytic fungi that attaches the fungus to the host (frequent in the Meliolaceae, black mildews).
Hypogeous	Located in the underground (soil)
Hypostroma	A compact mass of mostly intertwined hyphae below the true stroma of a fungus
Hypothecium	A layer of hyphae that is part of the wall forming below and around an ascoma, sometimes forming also a stipe that anchors the ascoma on the substrate.
Inoperculate	Dehiscence type of an ascus, not opening with a lid; inoperculate asci are mostly characterised by an apical pore, either delimited by an annulus or without distinct annulus (Fig. VIII.7j).
Integrated (conidiogenous cell)	The conidiogenous cell cannot be distinguished from the conidiophore cells (antonym: discrete).
Karyogamy	Mating of two nuclei (sexual reproduction)
Lenticular	Lens-shaped
Limoniform	Lemon-shaped
Loculus (pl. locula)	Chambers that can be equated to perithecia containing asci and ascospores. Typical of the so-called „Loculoascomycetes“ (Ascomycota with bitunicate asci).
Lunate	(half)moon-shaped
Macroconidia	Large, usually septate conidia produced by some fungi together with smaller, one-celled conidia called microconidia. Some strains of fungi that are able to produce both forms, however, may sometimes form only one type at a time.
Macronematous	Conidiophores that are morphologically different from the vegetative hyphae.

Margo	The margin of an apothecium
Mazaedium	A powdery mass of ascospores, disintegrating asci and paraphyses typical of some lichens
Meiosis	Cell division in sexually reproducing organisms, with reduction of the number of chromosomes in reproductive cells from diploid to haploid
Meiosporangium	A sporangium in which meiosis occurs (ascus, basidium)
Meiospore	A spore formed sexually (by meiosis)
Memnospore	Large, thick-walled, long-lived spore formed by the fungus with the aim to survive ecologically unfavourable conditions (s. Sect. XI).
Meristematic	See retrogressive
Merosporangium (pl. merosporangia)	Cylindrical small sporangium containing few spores, usually in a row, and developing on the tip of a swollen sporangiophore (Fig. VIII.1e)
Mesophilic	Optimal growth at temperatures of 10–40 °C
Metulae	Sterile cells subtending the phialides of some *Aspergillus* and *Penicillium* species (Fig. VIII.16, 18)
Microconidium	See macroconidium
Micronematous	A conidiophore that is not morphologically different from the vegetative hyphae. Often the conidiogenous cell originates directly from or is integrated in the hypha (Fig. VIII.11b, c).
Mitosis	The process by which the nucleus of a eukaryotic cell separates the chromosomes into two identical sets that are then segregated in two daughter nuclei. Asexual division.
Mitospore (mitotic spore)	Spore formed asexually by mitosis (conidium, sporangiospore)
Mono- (-blastic, -phialidic, - tretic)	Only one budding region on a conidiogenous cell
Mononematous	Conidiophores formed singly (antonym: synnematous)
Monoverticillate	Branching pattern in *Penicillium*: the phialides are formed directly on the conidiophore stipe (Fig. VIII.18a, IX.2a–b).

Napiform	Turnip-shaped
Navicular	Comma-shaped, boat-shaped
Obclavate	Inversely club-shaped
Oblong	With an elongated, often short-cylindrical shape
Obovoid	Inversely egg-shaped
Obpyriform	Inversely pear-shaped
Operculate	A unitunicate ascus opening with a distinct lid. Typical of Pezizomycotina.
Osmophilic	Growing at high osmotic pressure. Osmophilic fungi grow only on media rich in sucrose or salt.
Ostiolum	A distinct opening on the top of a perithecium, more rarely laterally, through which ascospores ooze and are thus dispersed (Fig. VIII.9d; VIII.7g).
Ovate	Egg-shaped
Paraphyses	Sterile structures arising among the asci in the hymenium. Paraphyses may be thread-like, clavate, lanceolate (lancet-shaped) or pointed. They may be branching, one-celled or septate.
Paraphysoids	Morphologically similar to the paraphyses, but typical of the bitunicate Ascomycota. Contrarily to paraphyses, however, they grow from the top of the fruit body and are thus intertwined with the apical part of the asci. In the lichens, they are often net-like and anastomosing. Sometimes also called pseudoparaphyses.
Percurrent	Growing continuously
Peridium	The outer layer of a fruit body. Depending on the fungal group, the peridium may have different types of texture (textura intricata, porrecta, epidermoidea, etc.).
Periphyses	Thread-like hyphal structures that line the inner wall of the ostiolum.
Perithecium (pl. perithecia)	Mostly spherical, elliptical or pear-shaped ascoma with an apical or more rarely lateral ostiolum, typical of unitunicate Ascomycota ("Pyrenomycetes") (Fig. VIII.7g)

Phialide, phialidic	Conidiogenous cell producing conidia in basipetal succession (the apical one is oldest) usually through one, more rarely several openings (Fig. VIII.5c; VIII.13e, g–i; VIII.14f)
Phragmosporous	Spore with several (at least more than one) transverse septa
Plectenchyma	A tissue formed by more or less tightly interwoven, thick-walled hyphae that at first sight appears tissue-like, with the hyphae having the appearance of long cells glued together
Pleomorphic	With more than one form (anamorph, teleomorph)
Poly- (blastic, phialidic, tretic)	More than one budding region on a conidiogenous cell (Fig. VIII.14d)
Prosenchyma	Plectenchymatous tissue in which the hyphae can still be recognised
Prototunicate	Asci characterised by a very thin ascus membrane that ruptures when the ascospores are ripe. Prototunicate asci are in general spherical, broadly club-shaped or (more rarely) cylindrical. They are usually contained in cleistothecia (Fig. VII.2.1d–f; VIII.5a, h; VIII.6a, c; VIII.7a, e).
Pseudomycelium	A mycelium-like structure composed of pseudohyphae, chains of easily disrupted fungal cells marked by constrictions rather than septa at the junctions (e.g. *Candida albicans*)
Pseudoparaphyses	See paraphysoids
Pseudoparenchyma	A tissue superficially resembling a plant parenchyma but composed of interwoven hyphae
Pseudostroma	A stromatic structure composed of fungal cells and remnants of host tissue
Pseudothecium (pl. pseudothecia)	Perithecium-like ascoma, in which asci are dispersed and not organized in a hymenium; a locule or cavity containing bitunicate asci
Psychrophilic	Growing optimally at 0–5 °C
Pycnidium (pl. pycnidia)	More or less closed structures, often resembling an ascoma, typical of some anamorphic fungi (Coelomycetes) (Fig. VIII.9d)
Pyrenomycetes	Unitunicate Ascomycota forming perithecia

Pyriform	Pear-shaped
Quadrangular	In section seemingly quadrangular (with 4 angles)
Reniform	Kidney-shaped
Reticulate	With a pattern of ridges, lines, cracks, or pigmentation on the surface
Retrogressive (conidiogenesis)	The tip of a hypha develops into a conidium; when the first conidium is formed, the second develops immediately below it and thus the conidiogenous hypha becomes shorter and shorter (Fig. VIII.6g–i). Called also meristematic conidiogenesis.
Rhizoid	Root-like structures (rhizoids) formed by hyphae and stolons at the point of contact with the substrate (mostly Mucorales; Fig. VIII.4a–b)
Rostrate	Forming a beak-like apical extension
Scleroplectenchym	A very densely packed structure of globose or angular, thick-walled cells
Sclerotium	Resting organ composed of a hard, compact mass or matrix of vegetative hyphae in or on which fruiting bodies are formed (Fig. VIII.12b, c).
Sculptured	With a rough surface (with spines, nets, streaks, or grains)
Semimacronematous	Conidiophores that are only slightly different from the vegetative hyphae (see also micronematous and macronematous).
Seta (pl. setae)	Bristle-like structure (Fig. VIII.8d)
Setae	Bristle; a stiff, generally thick-walled, multicellular hair, usually dark in colour (Fig. VIII.8d)
Sigmoid	Shaped like an S
Spatulate	Spoon-shaped or spatula-shaped
Sporangiole	Small, deciduous sporangium with only few spores, sometimes occurring along with larger sporangia; sporangioles have usually no or a very reduced columella (Fig. VIII.1b, d).

Sporangiophore	The structure bearing sporangia in the Mucorales (Fig VIII.4a)
Spore appendage	Slimy or cellular, often thread-like (more rarely bristle-like) appendage on a spore
Sporodochium (pl. sporodochia)	Erumpent, crowded cluster of conidiophores arising from the host surface in the form of a (sometimes stromatic) cushion; in culture sporodochia can hardly be distinguished from acervuli (Fig. VIII.14e).
Stolons	Runners; horizontal or vertical hyphae that run across the substrate and allow the fungus to colonise it; sometimes rhizoids are formed at the end of each stolon. See *Rhizopus* and *Absidia* (Fig. VIII.4d).
Striate	Marked with striae (stripes) (Fig. VIII.4e–f)
Stroma (pl. stromata)	1) A compact mass or matrix of vegetative fungal hyphae in or on which fruiting bodies are formed 2) Fruit body of bitunicate Ascomycota (only older literature, in this context not used any longer)
Subiculum (pl. subicula)	A thin layer or hyphae formed on the host surface, on which fruit bodies are formed
Suspensors (cells)	The hyphal structures that hold a zygospore (Fig. VIII.3e–g)
Sympodial	The conidiogenous cell grows sub-apically after each conidium formation, thus branching without a main axis but with several lateral branches (Fig. VIII.11a, f).
Synanamorph	An additional anamorph formed in addition to an already known one; the term is used when more than one anamorph is formed in a holomorph.
Synnema (pl. synnemata)	Large, erect structures composed by conidiophores fused together to form a strand resembling a stalk, with conidia at the end or on the lateral parts of it (Fig. VIII.15d).
Teleomorph	The sexual, reproductive stage of a fungus (asexual stage: anamorph)
Terverticillate	„Three-storied“ structural disposition in *Penicillium* conidiophores: the conidiogenous cells are disposed on sterile metulae, which in turn are disposed on short or long branches (rami) (Fig. VIII.18d–g, IX.2f).

Textura	Texture: structure of the peridium. In the specialised literature three different types are recognised (textura intricata, porrecta, epidermoidea).
Thallic	Conidia are formed by breaking of a hypha in several small units that differentiate into spores (s. also arthric) (Fig. VIII.10).
Thalloconidium	Conidium formed through thallic conidiogenesis (Fig. VIII.10c, g)
Thallus	The vegetative structures of a fungus (used almost synonymously to mycelium)
Thermophilic	Growing optimally at 50–60 °C, almost no growth at room temperature (20 °C)
Tretic	Blastic conidiogenesis, in which only the inner layers of the conidiogenous cell contribute to the production of the conidial wall; after conidium secession, a pore on the conidiogenous cell marks the point where the conidium was formed.
Triangular	In section with three angles
Truncate	With a flat base (Fig. VIII.17e)
Undeterminated	Conidiogenous cell growing further after conidium formation (antonym: determinated)
Unitunicate	Asci with a well-developed ascus membrane, which in light microscopy appears as a one-layered structure (Fig. VIII.7j). Unitunicate asci are frequently cylindrical or narrowly club-shaped and are mostly formed in perithecia or apothecia. Ascospores are ejected actively from the ascus through an apical pore (Fig. VII.2.1g), which can be variously shaped and sometimes provided by a distinct apical plug.
Vermiform	Worm-shaped, sigmoid
Verrucose	Covered with warts or wart-like projections
Xenospores	Thin-walled, small spores that germinate quickly and easily. They detach readily from the conidiogenous cells or the hyphae and are dispersed by wind or water. Their function is the rapid dispersal and distribution of propagules by a species.

Xerophilic	Growing and reproducing in conditions of low water availability (low water activity)
Zoophilic	(Of dermatophytes) colonising preferentially animals
Zoospore	Sexual or asexual, motile spore in the Protista and Chromista, usually provided with a whip
Zygosporangium	Fruit body of Zygomycota (Fig. VIII.3e–g)
Zygospore	The sexual spore of Zygomycota, formed in zygosporangia

XI. Ecology and physiology of fungi

1. Direct detection in environmental samples

Detection of fungi in environmental samples relies on direct and indirect methods.

Direct microscopy and histology (sections, swabs, clearing of tissues with chemical techniques) are fast, but they provide only limited information on the physiological and ecological activity of the organisms studied and do not allow any experimental approach.

The same applies to the use of **immunological** and, to a lesser extent, **molecular biology methods**. Immunofluorescence microscopy and DNA analysis (conventional PCR and real-time PCR) can provide evidence of colonisation of tissues by specific organisms and of their activity in a given ecosystem. Most of the time, however, the development of reliable molecular probes is only possible if the organisms to be studied are available in pure culture, which remains the obligate step before serious ecological and physiological work can be undertaken.

2. Isolation in pure culture

Molecular biology techniques can now be used routinely to detect and quantify microorganisms in a sample. Physiological and ecological investigations, however, can be best planned and carried out only if pure cultures of the organisms are available.

2.1 Dilution series

The first step in most ecological studies is the isolation of selected organisms from a substrate and the determination of their frequency. Dilution series are usually needed to allow counting of colony forming units in samples that usually contain large numbers of propagules. When dealing with fungi, the selective media used to prepare dilution series should contain antibiotics (e.g. tetracyclin, oxytetracyclin) to suppress bacterial growth. The antibiotics should be chosen to target both Gram positive and Gram negative bacteria, preference being given, if not absolutely needed otherwise, to substances not routinely used for human therapy.

2.2 Baiting

Some fungi are best attracted by selective substances on which the organisms normally grow best. For instance, the use of snakeskin has been shown to be excellent bait for Chytrids.

2.3 Single spore cultures

These are needed in particular for genetic studies and are useful to demonstrate teleomorph-anamorph connections, be it by conventional or molecular biology approaches. Single spore cultures can be easily prepared in any mycological laboratory, for example using a simple micromanipulator.

2.4 Surface sterilisation of plant tissues

This method is used to isolate organisms living endophytically in plant tissues. Detailed protocols can be found in the literature on fungal endophytes (e.g. Fröhlich et al. 2000; Petrini et al. 1982; Petrini 1986).

3. Adaptation of fungi to the ecosystem

Fungi are known to influence their environment and to induce significant physiological changes in it. In turn, the ecosystem greatly influences fungal growth, exerting a strong selective pressure. Correspondingly, the organisms have to develop strategies to ensure their survival. So-called stenoecious species are very much dependent on ecological changes, often not being able to cope with even small variations and thus to survive them, whereas euryoecious organisms can easily and quickly adapt to ecosystem alterations. Such organisms are in general highly competitive and able to survive in highly competitive situations.

3.1 Temperature requirements

Most fungi survive (but usually do not grow at) very low temperatures. In fact, most of them can be stored for long periods of time (in fact, years) at -20 °C or below. Ecologically, they can be classified according to their optimal growth temperatures.

- **Psychrophilic** fungi grow optimally between 10 and 15 °C, the maximal temperature allowing growth (and often survival) lies between 20 and 30 °C, whereas the minimal is approximately 0 °C. Example: *Monographella nivalis* (the wheat snow mould).

- **Mesophilic** fungi have an optimal growth at temperatures comprised between 18 and 24 °C. A small number of taxa (among them some important human pathogens) may be **eurythermic** (growing almost equally well over a large temperature range), as is the case of *Aspergillus fumigatus*, which grows between 3 and 55 °C.

- **Thermophilic** fungi grow optimally at temperatures over 40 °C, with maximal values between 45 and 60 °C. These fungi usually cannot grow (but they survive!) at temperatures below 20–30 °C. Example: *Thermoascus crustaceus*.

3.2 Light requirements and adaptation

Light, in particular short-wave light, may influence fungal growth, often stimulating the formation of asexual or sexual reproduction. This is thought to be dependent on the production of **mycosporins** by the fungal cells, which can influence positively or negatively spore development.

Some fungi show tropic reactions to light. For instance, species of *Pilobolus* (Zygomycota) react positively phototropically to light and shoot their spores consistently in the direction from which light comes. In alpine and arctic regions, or in other parts of the world where UV irradiation is high, fungi tend to develop thick-walled, strongly melanised spores that aim to protect the genetic material from UV-induced mutations. A prime example is given by alpine and arctic grass-inhabiting fungi such as *Pleospora* or *Phaeosphaeria* (Ascomycota).

4. Morphological and physiological adaptation

Ecologically, vegetative and reproductive growths fulfil different requirements. Vegetative growth aims at colonising a limited part of the ecosystem by a given individual, whereas asexual or sexual reproduction allows colonisation of even separate ecosystems by its descendants.

Ecologically, two types of spores can be distinguished (Gregory 1966).

Xenospores are usually thin-walled, small spores that germinate quickly and easily. They detach readily from the conidiogenous cells or the hyphae and are dispersed by wind or water. Their function is the rapid dispersal and distribution of propagules by a species.

Memnospores are resting spores. Their main objective is the protection of the genetic material during adverse ecological conditions. Memnospores are usually thick-walled, often melanised, rarely actively dispersed and often not easily detachable from the hyphae or the sporogenous cells. Germination often sets in only after special ecological events (e.g. heat shock, digestion by passage in the intestinal tract of ruminants).

XII. Moulds and health

1. Mycotoxins

1.1 Introduction

Mycotoxins are low-molecular weight, toxic secondary metabolites produced mainly by Ascomycota or, more rarely, Zygomycota growing on or in foodstuff and storage products. They usually do not elicit any immunological response by mammals exposed to or ingesting them. Most toxins are heat-resistant and may survive very high temperatures (e.g. half-life of ochratoxin: 12 min. at 270 °C). The most frequent and important mycotoxins are listed in Table XII.1.1.

Detailed information, including health hazards and risk reduction strategies is included e.g. in Cole & Cox (1981), Gravesen et al. (1994, 1999), Mücke & Lemmen (2005), Samson et al. (2004), Singh et al. (1991), Sinha & Bhatnagar (1998), Smith & Henderson (1991), and Smith & Moss (1985).

So far approximately 400 different mycotoxins have been described. They belong to 25 distinct chemical structural types and are produced by approximately 350 fungal species. Toxin production, however, is not always strictly species-specific but rather strain-specific, i.e. in some species only selected strains are toxin producers: in some cases, some strains of the same species may produce different toxins and others no toxins at all. The most prominent toxin producing fungi belong to the genera *Aspergillus*, *Fusarium*, and *Penicillium*.

Fungal toxins may contaminate foodstuff in three different ways:

- By **primary contamination** of the food ingredients by the toxin producing strain;
- By **secondary contamination** of the foodstuff by the toxin producing fungal strain;
- By a **"carry-over" effect**, whereby the toxins are taken up from contaminated products in the alimentary chain and transmitted to the next consumer.

Mycotoxin production is heavily influenced by substrate composition, temperature, pH, water activity and other micro-environmental factors: for instance, aflatoxin production by *A. flavus* is dependent on the concentration of Zn^{2+} ions in the substrate. Mycotoxin production takes place preferentially at high water activity values; some toxins (e.g. patulin and trichothecenes) are produced at low temperatures (e.g. cool storage conditions); finally, some fungi are known to form toxins only under laboratory conditions and rarely or never in their original habitat (e.g. sterigmatocystin production by *Emericella nidulans* and *Chaetomium*).

Tab. XII.1.1: Most frequent and economically important mycotoxins.

Fungal species	Mycotoxin
Aspergillus parasiticus, *A. flavus*	Aflatoxins
Aspergillus versicolor	Sterigmatocystins
A. alutaceus, *A. niger*	Tremorgens; Cytochalasins; Malformin C
Aspergillus ochraceus, *Penicillium verrucosum*	Ochratoxins
A. clavatus, *Penicillium citrinum*, *P. expansum*, *Penicillium* spp.	Patulin, Citrinin, Rubratoxins
Eupenicillium spp., *Byssochlamys* spp., *Paecilomyces* spp.	Patulin
Claviceps spp.	Tremorgene (Indole derivatives)
Gibberella avenacea, *G. zeae*, *Fusarium culmorum*, *F. graminearum*	Trichothecenes, Zearalenone
Phoma exigua, *P. herbarum*, *Xylaria* spp.	Cytochalasines; Brefeldin A
Fusarium avenaceum, *F. zeae*, *Fusarium* spp.	Zearalenone; FZ-Toxin; Vomitoxin
Pithomyces chartarum	Sporidesmins
Alternaria alternata	Tenuazonic acid
Curvularia lunata	Brefeldin A
Stachybotrys spp.	Trichothecenes, atranones
Verticillium spp.	Dimeric sporidesmins

Mycotoxins may be produced already in the field, in fodder or in the stored material used for food production. Some aliments that may contain mycotoxins through contamination of the material used to manufacture human and animal food are listed in Tab. XII.1.2. Food safety measures are used to avoid or limit mycotoxin contamination of marketed products.

Tab. XII.1.2: Foodstuff that may often be contaminated with mycotoxins (from Weber 1993)

Cereals	Nuts*, berries	Animal products	Fruit
Rice	Peanuts	Milk	Tomato juice
Corn	Walnuts	Cheese	Apple juice
Rye	Hazelnuts	Egg products	Wine
Wheat	Almonds	Liver	Fruit juices
Soy	Coffee beans	Kidney	

*mostly aflatoxins

1.2 Mycotoxin detection

Mycotoxin detection relies on biological and (increasingly) chemical methods.

Biological tests include the rabbit skin test (trichothecenes and satrotoxins), the chicken embryo test (scirpenols) and the pollen test. Feeding experiments, in former times standard testing procedures, have almost completely been replaced by chemical and other *in vitro* investigations. Biological tests, in general, are gradually abandoned, because they allow detection of only few toxins and also their sensitivity is limited.

Chemical analysis is very reliable and sensitive, the detection limits being quite low. The methods used (Thin-layer chromatography, HPLC, gas chromatography, ELISA) are described in detail by Truckess & Pholand (2000). Egmond (2000), Frisvad & Thrane (1987, 1993), Frisvad et al. (1989), and Samson & Pitt (2000) report the methods used for mycotoxin detection in selected fungal genera.

2. Health hazards caused by mycotoxicoses

Poisonings of humans and animals caused by mycotoxins are termed mycotoxicoses. Mycotoxicosis is mainly the result of the inhalation or ingestion of toxic volatile compounds (mVOC: microbial volatile organic compounds) or spores released in the environment or present in the ingested food. The most relevant mycotoxicoses are listed in Tab. XII.2.1.

Tab. XII.2.1: Some mycotoxicoses and their causal agents (from Mücke & Lemmen 2005; Egmond 2000)

Toxicosis	Causal agent	Toxins
Fusarium-Mycotoxicosis (e.g. „Hole in the head syndrome“, 1989)	*Fusarium* spp.	Fusarin C, Zearalenon, Trichothecenes, Fumonisin
Ochratoxicosis („Balkan endemic Nephropathy“, 1952)	*Penicillium verrucosum* (*Aspergillus alutaceus*)	Ochratoxin A
Aflatoxicosis (Turkey X disease, 1960)	*Aspergillus flavus*, *A. parasiticus*	Aflatoxins (mainly B1, B2, G1, G2)
Cardiac beriberi (1890)	*Penicillium citreonigrum*	Citreoviridin
Ergot poisoning („holy fire“)	*Claviceps* spp., grass endophytes	Ergot alkaloids

Acute intoxications are usually the consequence of the ingestion of very mouldy foodstuff containing large amounts of toxins; the chronic intake of small amounts of mycotoxins, on the other hand, may also have serious and sometimes fatal outcomes, some of the toxins being carcinogenic or causing severe pathological changes. Aflatoxicosis is a prime example: in those regions where the aflatoxin load is very high the risk of liver cirrhosis and liver carcinoma is dramatically increased (Mücke & Lemmen 2004).

Inhalation of mycotoxins or spores can cause severe allergic reactions and asthma in debilitated and predisposed subjects (immunosuppressed or already sick people, people with known allergies to other substances). These allergic reactions may often develop into mycoses in debilitated patients (e.g. aspergillosis caused by *A. fumigatus*).

Allergies in people exposed through their profession to high loads of fungal allergens are quite common. Allergic reactions caused by fungi have been reported from malt and cheese factories, bakeries, mills, and composting facilities, all sites where high loads of fungal spores may be present, be it as contaminants (as in the case of mouldy barley in the malt factories) or as agents used in the production of selected food types (e.g. *Penicillium candidum* in soft cheese). High fungal loads can also be present in damp buildings that have not been properly restored, particularly after serious water damage (Gravesen et al. 1999; Mücke & Lemmen 2004; Nielsen et al. 1998a, b, 1999, 2000; Samson et al. 1994).

3. Mycoses

Thanks to their ability to produce a large variety of enzymes, fungi are able to colonise almost all organic substrates. In this respect, animal tissues are no exception. Fungi may colonise their animal hosts by growing in the tissue without causing any harm, or infect them by invading the tissue and eliciting local cellular injury. The borderline between colonisation and infection is very thin and often only minimal alterations of the immunological conditions of the host are needed for the fungal parasite to switch from the commensal to the pathogenic state.

The medically interested reader may refer to Anaissie et al. (2009).

3.1 Hypersensitivity reactions

Colonisation of human tissues by fungi is not uncommon. For instance, the yeast *Candida albicans* is a common commensal on human skin and mucosae (Calderone 2002). Other moulds, in particular air-borne organisms, are able to colonise human and animal tissues, often causing hypersensitivity reactions in the airways of predisposed individuals, which may manifest themselves as hypersensitivity pneumonia (also called extrinsic allergic alveolitis).

Tab. XII.3.1. Examples of fungus-related hypersensitivity reactions

Disease	Major antigen	Exposure
Mushroom worker's lung	Thermophilic actinomycetes	Mushroom spores, mushroom compost
Malt worker's lung	*Aspergillus clavatus*, *Faenia rectivirgula*	Mouldy barley
Cheese worker's lung	*Penicillium casei* / *P. candidum*	Cheese mould
Paprika splitter's lung	*Mucor stolonifer*	Paprika dust
Compost lung	*Aspergillus* spp.	Compost
Peat moss worker's lung	*Monocillium* spp., *Penicillium citreonigrum*	Peat moss
Maple bark stripper's lung	*Cryptostroma corticale*	Mouldy wood bark
Wood pulp worker's lung	*Alternaria* spp.	Mouldy wood pulp

Extrinsic allergic alveolitis has been defined as „the result of an over zealous immune response to inhaled antigens" (Warren 1977). A typical form of hypersensitivity pneumonia is the so-called farmer's lung, a consequence of the inhalational exposure to thermophylic actinomycetes and various *Aspergillus* species. Farmer's lung is comparatively common in cold, damp climates in late winter and early spring and its incidence is very variable (USA: 8–540 cases per 100,000 at-risk individuals per year). Mortality rate from farmer's lung is also variable (0–20%, according to several reports), with death occurring within 5 years after diagnosis. Systemic corticosteroid administration and avoidance measures are the treatment of choice. Other forms of hypersensitivity reactions to fungal material occur in people who work in places where the levels of mould spores are very high; some examples are provided in Tab. XII.3.1.

3.2 Mycoses

The incidence of mycoses is continuously rising (Pfaller 2010), particularly in immunocompromised people. For example, the annual number of US cases of sepsis caused by fungi has increased by more than 200% between 1979 and 2000 (Martin et al. 2003). While the normal, immunocompetent host may be infected, albeit only rarely, by only a handful of species, more than 300 fungal taxa, mainly saprobes, are known as opportunistic pathogens of immunocompromised patients. Yeasts, in particular species of *Candida* (e.g. *C. albicans*, *C. krusei*, *C. glabrata*, *C. parapsilosis*), some *Aspergillus* species and the dermatophytes are the most common agents.

Mycoses can be systemic (invasive) or localised. Overall, only 25 fungal species are considered to be obligate systemic pathogens, while all others are either obligate cutaneous or subcutaneous parasites (approx. 40 species), or opportunistic infectious agents.

Primary mycoses are fungal infections in normal hosts, whereas secondary mycoses are typical primarily of immunocompromised patients. Opportunistic mycoses are caused by fungi that are not necessarily pathogenic but can occasionally infect at-risk people such as individuals undergoing hematopoietic stem cell or solid organ transplantation, having major surgery, AIDS and oncology patients, patients under immunosuppressive therapy, people in advanced age, and premature birth infants. Some examples of mycoses are listed in Tab. XII.3.2.

Fungi can enter their host through a number of portals, the most important being airways, mouth, skin, and urogenital tract. Wounds, including those related to surgical interventions, may also be routes of infection for many opportunistic fungi.

Fungal pathogenicity is still poorly unknown. Many fungal species are thermotolerant or microaerobic, or both, thus being able to survive in the host tissues. In addition, they are also fit to withstand host defences by producing several pathogenicity factors such as adhesins, enzymes (elastases, keratinases, proteases, phospholipases, phenoloxidases), toxins (haemolysin, endotoxin, canditoxin), siderophores, or melanin; dimorphism, encapsulation and resistance against phagocytosis are additional elements that allow fungi to successfully infect a comparatively large range of at-risk hosts.

Tab. XII.3.2. Examples of mycoses

Opportunistic	e.g. *Alternaria* spp., *Aspergillus* spp., *Acremonium* spp., *Schizophyllum* spp.,
Systemic	
Aspergillosis	*Aspergillus fumigatus*
Cryptococcosis	*Cryptococcus neoformans*
Coccidiomycosis	*Coccidioides inmitis*
Subcutaneous	
Mycetoma	Several agents
Sporotrichosis	*Sporothix schenckii*
Dermatophytoses (= Dermatomycoses)	Dermatophytes
e.g. tinea, onychomycosis	e.g. *Trichophyton rubrum*, *T. mentagrophytes*
Candidiasis	*Candida albicans*

Treatment of systemic (invasive) mycoses is difficult and often requires a combination of therapeutic approaches. The antifungal armamentarium includes old but still very effective drugs such as amphotericin B and flucytosin, the azoles (e.g. fluconazole, voriconazole, posaconazole and new generation azoles) and the echinocandins. A comprehensive review is found in Thompson et al. (2009).

A special group of fungi causing mycoses are the dermatophytes. This group of organisms, mainly belonging to the Ascomycota, are probably the most frequently observed fungal organisms in medicine. They include approximately 25 pathogenic species and they may cause, worldwide, healthcare costs of over $ 500,000,000 p.a. Their identification is comparatively easy and is carried out by morphology and molecular biology; MALDI-TOF is a promising technique that is slowly finding its application also within this fungal group.

Dermatophytes are roughly divided in three groups. Anthropophilic species (e.g. *Trichophyton rubrum*, *Epidermophyton*) are associated with humans only and the transmission occurs through contaminated objects; zoophilic species (for instance *Microsporum canis*: cats and dogs; *M. nanum*: swine; *Trichophyton verrucosum*: horse and swine) are normally associated with animals. A direct transmission to humans is possible by close contact with an infected animal. Finally, geophilic organisms (e.g. *M. gypseum*) are usually found in soil and may infect humans by direct exposure. Griseofulvin, tinactin, clotrimazole, miconazole, ketoconazole, itraconazole, and terbinafine are the antifungal agents used for the treatment of dermatophytes.

XIII. References

Aa, van der, H. A. (1973). Studies in *Phyllosticta* I. – Studies in Mycology **5**: 1–110.

Aa, van der, H. A. & S. Vanev (Hrsg. A. Aptroot, R. C. Summerbell, & G. J. Verkley, 2002). A revision of the species described in *Phyllosticta*. – Centraalbureau voor Schimmelcultures, Utrecht, 510 pp.

Abbott, S. P., L. Sigler & R. S. Currah (1998). *Microascus brevicaulis* sp. nov., the teleomorph of *Scopulariopsis brevicaulis*, supports placement of *Scopulariopsis* with the *Microascaceae*. – Mycologia **90**: 297–302.

Ames, L. M. (1963). A monograph of the *Chaetomiaceae*. – U. S. Army Res. Dev. Ser. 2, 125 pp.

Anaissie, E. J., M. R. McGinnis & M. A. Pfaller (2009). Clinical Mycology. – Elsevier Health Sciences, 688 pp.

Apinis, A. E. & H. W. O. Eggins (1966). *Thermomyces ibadanensis* from oil palm kernel stacks in Nigeria. – Trans. Br. Mycol. Soc. **49**: 629–632.

Arzanlou M., J. Z. Groenewald, W. Gams, U. Braun, H.-D. Shin & P. W. Crous (2007). Phylogenetic and morphotaxonomic revision of *Ramichloridium* and allied genera. – Studies in Mycology **58**: 57–93.

Aveskamp, M., H. de Gruyter, J. Woudenberg, G. Verkeley & P. W. Crous (2010). Highlights of the *Didymellaceae*: a polyphasic approach to characterise *Phoma* and related pleosporalean genera. – Studies in Mycology **65**: 1–60.

Balajee, S. A., J. Houbraken, P. E. Verweij, S. B. Hong, T. Yaghuchi, J. Varga et al. (2007). *Aspergillus* species identification in the clinical setting. – Studies in Mycology **59**: 39–46.

Barkham, T. & M. B. Taylor (2002). Sniffing bacterial cultures on agar plates: a useful tool or a safety hazard? – J. Clin. Microbiol. **40** (10): 3877.

Barron, G. L. (1968). The genera of hyphomycetes from soil. – Williams & Wilkins, Baltimore, 364 pp.

Barron, G. L., R. F. Cain & J. C. Gilman (1961). The genus *Microascus*. – Can. J. Bot. **39**: 1609–1631.

Benjamin, C. R. & C. W. Hesseltine (1957). The genus *Actinomucor*. – Mycologia **49**: 240–249.

Benjamin, C. R. & C. W. Hesseltine (1959). Studies on the genus *Phycomyces*. – Mycologia **51**: 751–771.

Bensch, K., U. Braun, J. Z. Groenewald & P. W. Crous (2012). The genus *Cladosporium*. – Studies in Mycology **72**: 1–401.

Bensch, K., J. Z. Groenewald, J. Dijksterhuis, M. Starink-Willemse, B. Andersen et al. (2010). Species and ecological diversity within the *Cladosporium cladosporioides* complex (*Davidiellaceae*, *Capnodiales*). – Studies in Mycology **67**: 1–94.

Berbee, M. L. & J. W. Taylor (1995). From 18S ribosomal sequence data to evolution of morphology among the fungi. – Can. J. Bot. **73** (Supp. 1): S677–S683.

Beyrde, R. J. W. & H. J. Willetts (1977). The brown rot fungi of fruit. Their biology and control. – Pergamon Press, 171 pp.

Bille, J. & G. Schär (2001). Medical mycology and the diagnostic laboratory. – Labolife, Schweiz. Ges. Mikrobiologie **2** (1): 5–8.

Bissett, J. (1979). Coelomycetes in Liliales: the genus *Phyllosticta*. – Can. J. Bot. **57**: 2082–2095.

Bissett, J. (1984). A revision of the genus *Trichoderma*. I. Section *Longibrachiatum* sect. nov. – Can. J. Bot. **62**: 924–931.

Bissett, J. (1986a). A note on the typification of *Guignardia*. – Mycotaxon **25**: 519–522.
Bissett, J. (1986b). *Discochora yuccae* sp. nov. with *Phyllosticta* and *Leptodothiorella* synanamorphs. – Can J. Bot. **64**: 1720–1726.
Bissett, J. (1991a). A revision of the genus *Trichoderma*. II. Infrageneric classification. – Can. J. Bot. **69**: 2357–2372.
Bissett, J. (1991b). A revision of the genus *Trichoderma*. III. Section *Pachybasium*. – Can. J. Bot. **69**: 2373–2417.
Bissett, J. (1991c). A revision of the genus *Trichoderma*. IV. Additional notes on section *Longibrachiatum*. – Can. J. Bot. **69**: 2418–2420.
Bissett, J. & M. E. Palm (1989). Species of *Phyllosticta* on conifers. – Can. J. Bot. **67**: 3378–3385.
Blaser, P. (1976). Taxonomische und physiologische Untersuchungen über die Gattung *Eurotium* Link : Fr. – Sydowia **28**: 1–49.
Boedijn, K. B. & J. Reitsman (1950). Notes on the genus *Cylindrocladium*. – Reinwardtia **1**: 51–60.
Boerema, G. H. (1976). The *Phoma* species studied in culture by Dr. R. W. G. Dennis. – Trans. Br. Mycol. Soc. **67**: 289–319.
Boerema, G. H. (1993). Contributions towards a monograph of *Phoma* (Coelomycetes). II. Section *Peyronellaea*. – Persoonia **15**: 197–221.
Boerema, G. H. & J. De Gruyter (1998). Contributions towards a monograph of *Phoma* (Coelomycetes) – VII – Section *Sclerophomella*: Taxa with thick-walled pseudoparenchymatous pycnidia. – Persoonia **17**: 81–95.
Boerema, G. H., J. De Gruyter & H. A. van Kesteren (1994). Contributions towards a monograph of *Phoma* (Coelomycetes) – III: 1. Section *Plenodomus*: Taxa often with a *Leptosphaeria* teleomorph. – Persoonia **15**: 431–487.
Boerema, G. H., J. De Gruyter & M. E. Noordeloos (1997). Contributions towards a monograph of *Phoma* (Coelomycetes) – IV. Section *Heterospora*: taxa with large sized conidial dimorphs, in vivo sometimes as *Stagonospora* synanamorphs. – Persoonia **16**: 335–371.
Boerema, G. H. & M. M. J. Dorenbosch (1973). The *Phoma* and *Ascochyta* species described by Wollenweber & Hochapfel in their study on fruit-rotting. – Studies in Mycology **3**: 1–50.
Boerema, G. H., M. M. J. Dorenbosch & H. A. van Kesteren (1977). Remarks on species of *Phoma* referred to *Peyronellaea*, V. – Kew Bull. **31**: 533–544.
Boerema, G. H., W. M. Loerakker & M. E. C. Hamers (1996). Contributions towards a monograph of *Phoma* (Coelomycetes) – III: 2. Misapplications of the type species name and the generic synonyms of section *Plenodomus* (excluded species). – Persoonia **16**: 141–190.
Boerema, G. H., H. A. van Kesteren & W. M. Loerakker (1981).Notes on *Phoma*. – Trans. Br. Mycol. Soc. **77**: 61–74.
Booth, C. (1959). Studies of Pyrenomycetes: IV. *Nectria* (Part I). – Mycological Papers **73**: 1–115.
Booth, C. (1966). The genus *Cylindrocarpon*. – Mycological Papers **104**: 1–56.
Booth, C. (1971a). Fungal culture media. – In: Norris, J.R. & D.W. Ribbons (Hrsg.). Methods in Microbiology, Band IV. London Academic Press, London: 49–94.
Booth, C. (1971b). The genus *Fusarium*. – Commonwealth Mycological Institute, Kew Surrey, England, 237 pp.
Booth, C. (1977). *Fusarium* Laboratory guide to the identification of the major species. – Commonwealth Mycological Institute, Kew Surrey, England, 58 pp.

Brunner, F. & O. Petrini (1992) Taxonomy of some *Xylaria* species and xylariaceous endophytes by isozyme electrophoresis. – Mycological Research **96**: 723–733.
Bruns, T. D., T. J. White & J. W. Taylor (1991). Fungal molecular systematics. – Annu. Rev. Ecol. Syst. **22**: 525–564.
Buchanan, P. K. (1987). A reappraisal of *Ascochyta* and *Ascochytella* (Coelomycetes). – MycologicalPapers **156**: 1–83.
Burgess, L. W., C. M. Liddell & B. A. Summerell (1988). Laboratory manual for *Fusarium* research. – University of Sydney, Sydney, 2nd ed.
CABI databases. – http://www.indexfungorum.org/
Calderone, R. A. (ed.) (2002). *Candida* and candidiasis. – ASM Press, Washington, D.C.
Cannon, P. F. & D. L. Hawksworth (1982). A re-evaluation of *Melanospora* Corda and similar pyrenomycetes, with a revision of the British species. – J. Linn. Soc. (Bot.) **84**: 115–160.
Carmichael, J. W. (1962). *Chrysosporium* and some other aleurosporic hyphomycetes. – Can. J. Bot. **40**: 1137–1173.
Castellani, E. & G. Germano (1977). Le Stagonosporae graminicole. – Ann. Fac. Sci. Agrar. Univ. Torino **10**: 1–135.
Cavalier-Smith, T. (1981). Eukaryote kingdoms: seven or nine? – Biosystems **14** (3-4): 461–81.
Cavalier-Smith, T. (1993). Kingdom protozoa and its 18 phyla. – Microbiol. Rev. **57** (4): 953–94.
Cavalier-Smith, T. (1998). A revised six-kingdom system of life. – Biol. Rev. Camb. Philos. Soc. **73** (3): 203–66.
Cavalier-Smith, T. (2004). Only six kingdoms of life. – Proc. Biol. Sci. **22**: 1251–62.
Cavalier-Smith, T. (2006). Rooting the tree of life by transition analyses. – Biol. Direct. **1**: 19.
CDC (Center for Disease Control and Prevention). 2009. Biosafety in microbiological and biomedical laboratories (BMBL). 5th edition. 415 pp. – http://www.cdc.gov/biosafety/publications/bmbl5/BMBL.pdf
Cerdá-Olmeda, E. & E. D. Lipson (1987). *Phycomyces*. – Cold Spring Harbor Laboratory Press, Cold Spring Harbor, New York, USA, 430 pp.
Checa, J., J. M. Barrasa, M. Moreno, F. Fort, & J. Guarro (1988). The genus *Coniochaeta* (Sacc.) Cooke (*Coniochaetaceae*, Ascomycotina) in Spain. – Cryptogamie Mycologie **9** (1): 1–34.
Chaverri, P., L. A. Castelbury, G. J. Samuels & D. M. Geiser (2003). Multilocus phylogenetic structure within the *Trichoderma harzianum/Hypocrea lixii* complex. – Mol. Phylogenet. Evol. **27**: 302–313.
Chaverri, P. & G. J. Samuels (2003). *Hypocrea/Trichoderma* (*Ascomycota*, *Hypocreales*, *Hypocreaceae*): species with green ascospores. – Studies in Mycology **48**: 1–116.
Chelkowski, J. (Ed., 1989). *Fusarium*: Mycotoxins, taxonomy and pathogenicity. – Elsevier Science Publishers B. V., Amsterdam, Oxford, New York, Tokyo, 492 pp.
Cleveland, W. S. (1994). The elements of graphing data. – Hobart Press, Summit, New Jersey, USA. 297 pp.
CMI Descriptions of pathogenic Fungi and Bacteria No. 616: *Sclerotinia fructicola.*
CMI Descriptions of pathogenic Fungi and Bacteria No. 617: *Sclerotinia fructigena.*
CMI Descriptions of pathogenic Fungi and Bacteria No. 618: *Sclerotinia homoeocarpa.*
CMI Descriptions of pathogenic Fungi and Bacteria No. 619: *Sclerotinia laxa.*
Cole, R. J. & R. H. Cox (1981). Handbook of toxic fungal metabolites. – Academic Press, New York, 937 pp.

Cole, G. T. & B. Kendrick (1973). Taxonomic studies of *Phialophora.* – Mycologia **65**: 661–688.

Colwell, R. R. (1969). Numerical taxonomy of the flexibacteria. – J. Gen. Microbiol. **58**: 207–215.

Colwell, R. R. (1970). Polyphasic taxonomy of the genus *Vibrio*: numerical taxonomy of *Vibrio cholerae*, *Vibrio parahaemolyticus*, and related *Vibrio* species. – J. Bacteriol. **104**: 410–433.

Cooney, D. G. & R. Emerson (1965). Thermophilic fungi. – San Francisco, 188 pp.

Corbaz, R. (1957). Recherches sur le genre *Didymella.* – Phytopath. Z. **28**: 375–414.

Corlett, M. (1981). A taxonomic survey of some species of *Didymella* and *Didymella*-like species. – Can. J. Bot. **59**: 2016–2042.

Crous, P. W. (2002). Taxonomy and pathology of *Cylindrocladium* (*Calonectria*) and allied genera. – APS Press, St. Paul, Minnesota, 294 pp.

Crous, P. W., U. Braun, K. Schubert & J. Z. Groenewald (2007). The genus *Cladosporium* and similar dematiaceous hyphomycetes. – Studies in Mycology **58**: 1–253.

Crous, P. W., W. Gams, J.A. Stalpers, V. Robert & G. Stegehuis (2004). MycoBank: an online initiative to launch mycology into the 21st century. – Studies in Mycology **50**: 19–22.

Crous, P. W., G. J. M. Verkeley, J. Z. Groenewald & R. A. Samson (2009). CBS Laboratory Manual Series 1: Fungal biodiversity. – Centraalbureau voor Schimmelcultures, Utrecht, 270 pp.

Crous, P. W. & M. J. Wingfield (1994). A monograph of *Cylindrocladium*, including anamorphs of *Calonectria.* – Mycotaxon **51**: 341–435.

Dabinett, P. E. & A. M. Wellman (1973). Numerical taxonomy of the genus *Rhizopus.* – Can. J. Bot. **51**: 2053–2064.

Dakin, J. C. & A. C. Stolk (1968). *Moniliella acetoabutans*: some further characteristics and industrial significance. – J. Fd. Technol. **3**: 49–53.

Damm, U., P. F. Cannon & P. W. Crous (2012). *Colletrotrichum*: complex species or species complex? – Studies in Mycology **73**: 1–213.

David, J. C. (1997). A contribution to the systematics of *Cladosporium.* – Mycological Papers **172**: 1–157.

De Bertoldi, M. (1976). New species of *Humicola*: an approach to genetic and biochemical classification. – Can. J. Bot. **54**: 27–55.

De Gruyter, J. & M. E. Noordeloos (1992). Contributions towards a monograph of *Phoma* (Coelomycetes) – I: 1. Section *Phoma*: taxa with very small conidia in vitro. – Persoonia **15**: 71–92.

De Gruyter, J., M. E. Noordeloos & G. H. Boerema (1993). Contributions towards a monograph of *Phoma* (Coelomycetes) – I: 2. Section *Phoma*: Additional taxa with very small conidia and taxa with conidia up to 7 µm long. – Persoonia **15**: 369–400.

De Gruyter, J., M. E. Noordeloos & G. H. Boerema (1998). Contributions towards a monograph of *Phoma* (Coelomycetes) – I: 3. Section *Phoma*: Taxa with conidia longer than 7 µm. – Persoonia **16**: 471–490.

De Hoog, G. S. (1977). *Rhinocladiella* and allied genera. – Studies in Mycology **15**: 1–140.

De Hoog, G. S. (1979). The black yeasts, II: *Moniliella* and allied genera. – Studies in Mycology **19**: 1–36.

De Hoog, G. S. (Hrsg., 1999). Ecology and evolution of black yeasts and their relatives. – Studies in Mycology **43**: 1–208.

De Hoog, G. S. & M. Grube (Hrsg., 2008). Black fungal extremes. – Studies in Mycology **61**: 1–194.

De Hoog, G. S. & E. J. Hermanides-Nijhof (1977). Survey of black yeasts and allied fungi. – Studies in Mycology **15**: 178–222.

De Hoog, G. S., J. Guarro, J. Gené & M. J. Figueras (2000). Atlas of clinical fungi. – Centraalbureau voor Schimmelcultures, Baarn, 2nd ed., 1126 pp.

De Hoog, G. S., M. Th. Smith & E. Guého (1986). A revision of the genus *Geotrichum* and its teleomorphs. – Studies in Mycology **29**: 1–131.

De Respinis, S., G. Vogel, C. Benagli, M. Tonolla, O. Petrini & G. J. Samuels (2010). MALDI-TOF MS of *Trichoderma*: A model system for the identification of microfungi. Mycol. Prog. **9** (1): 79–100. DOI 10.1007/s11557-009-0621-5.

De Ruiter, G. A., S. H. W. Notermans & F. M. Rombouts (1993). Review: new methods in food mycology. – Trends Food Sci. Technol. **4**: 91–97.

Degenkolb, T., R. Dieckmann, K. F. Nielsen, T. Gräfenhan, C. Theis, D. Zafari, P. Chaverri, A. Ismaiel, H. Brückner, H. von Döhren, U. Thrane, O. Petrini & G. J. Samuels (2008). The *Trichoderma brevicompactum* clade: a separate lineage with new species, new peptaibiotics, and mycotoxins. – Mycol. Prog. **7**: 177–219.

Desjardin, A. E. (2006). Fusarium mycotoxins, chemistry, genetics, and biology. – APS Press. St. Paul, Minnesota. USA, 260 pp.

DiCosmo, F., T. R. Nag Raj & B. Kendrick (1983). Prodromus for a revision of the *Phacidiaceae* and related anamorphs. – Can. J. Bot. **61**: 31–44.

DiCosmo, F., T. R. Nag Raj & B. Kendrick (1984). A revision of the *Phacidiaceae* and related anamorphs. – Mycotaxon **21**: 1–234.

Dörge, T., J. M. Carsensen & J. C. Frisvad (2000). Direct identification of pure *Penicillium* species using image analysis. – J. Microbiol. Meth. **41**: 121–133.

Doguet, G. (1955). Le genre *Melanospora*. – Botaniste **39**: 1–313.

Doi, Y. (1972). Revision of the *Hypocreales* with cultural observations IV. The genus *Hypocrea* and its allies in Japan. – Bull. Natn. Sci. Mus. Tokyo **15**: 649–751.

Doi, Y. & K. Yamatoya (1989). *Hypocrea pallida* and its allies (*Hypocreaceae*). – Memoirs of the New York Botanical Garden **49**: 233–242.

Domsch, K. H., W. Gams & T.-H. Anderson (2007). Compendium of soil fungi. – 2nd ed. revised by W. Gams. IHV-Verlag, Eching, 672 pp.

Dorenbosch, M. M. J. (1970).Key to nine ubiquitous soil-borne *Phoma*-like fungi. – Persoonia **6**: 1–14.

Dreyfuss, M. (1975). Taxonomische Untersuchungen innerhalb der Gattung *Chaetomium* Kunze. – Sydowia **28**: 50–133.

Druzhinina, I. S., A. G. Kopchinskiy, M. Komon, J. Bissett, G. Szakacs, C. P. Kubicek (2005). An oligonucleotide barcode for species identification in *Trichoderma* and *Hypocrea*. – Fungal Genet. Biol. **42** (10): 813–28.

Druzhinina, I. S., A. G. Kopchinskiy & C. P. Kubicek (2006). The first 100 *Trichoderma* species characterized by molecular data. – Mycosciences **47**: 55–64.

Druzhinina, I. S., M. Komon-Zelazowska, L. Kredics, L. Hatvani, Z. Antal, T. Belayneh & C. P. Kubicek (2008). Alternative reproductive strategies of *Hypocrea orientalis* and genetically close but clonal *Trichoderma longibrachiatum*, both capable of causing invasive mycoses of humans. – Microbiology **154**: 3447–3459.

Egmond, H. P. van (2000). Mycotoxins: detection, reference material and regulation. – In: Samson, R. A., E. S. Hoekstra, J. C. Frisvad & O. Filtenborg (Eds). Introduction to food- and air borne fungi. Centraalbureau voor Schimmelcultures: 332–338.

Ellis, M. B. (1960). Dematiaceous Hyphomycetes: 1. – Mycological Papers **76**: 1–36.

Ellis, M. B. (1966). Dematiaceous hyphomycetes. VII: *Curvularia*, *Brachysporium*etc. – Mycological Papers **106**: 1–57.

Ellis, M. B. (1971). Dematiaceous hyphomycetes. – Commonwealth Mycological Institute, Kew, 608 pp.

Ellis, M. B. (1976). More dematiaceous hyphomycetes. – Commonwealth Mycological Institute, Kew, Surrey, England, 507 pp.

Ellis, M. B. & J. P. Ellis (1988). Microfungi on miscellaneous substrates. – Croom Helm, London & Sydney, 244 pp.

Eriksson, O. (1995). DNA and ascomycete systematics. – Can. J. Bot. **73** (Suppl. 1): S784–S789.

Fassatiová, O. (1967). Notes on the genus *Humicola*. – Česká Mykol. **21**: 78–89.

Forrest G. N., K. Mankes, M. A. Jabra-Rizk, E. Weekes, J. K. Johnson, D. P. Lincalis et al. (2006). Peptide nucleic acid fluorescence in situ hybridization-based identification of *Candida albicans* and its impact on mortality and antifungal therapy costs. – J. Clin. Microbiol. **44** (9): 3381–3383.

Frisvad, J. C., P. D. Bridge & D. K. Arora (Hrsg., 1998). Chemical fungal taxonomy. – Marcel Dekker Inc., New York, USA, 398 pp.

Frisvad, J. C. & O. Filtenborg (1989). Terverticillate *Penicillia*: chemotaxonomy and mycotoxin production. – Mycologia **81**: 837–861.

Frisvad, J. C., O. Filtenborg, R. A. Samson & A. C. Stolk (1990). Chemotaxonomy of the genus *Talaromyces*. – Ant. van Leeuwenhoek **57**: 179–189.

Frisvad, J. C., O. Filtenborg & U. Thrane (1989). Analysis and screening for mycotoxins and other secondary metabolites in fungal cultures by thin-layer chromatography and high-performance liquid chromatography. – Arch. Environ. Contam. Toxicol. **18**: 331–335.

Frisvad, J. C. & R. A. Samson (2000). *Neopetromyces* gen. nov. and an overview of teleomorphs of *Aspergillus* subgenus *Circumdati*. – Studies in Mycology **45**: 201–207.

Frisvad, J. C. & U. Thrane (1987). Standardized high-performance liquid chromatography of 182 mycotoxins and other fungal metabolites based on alkylphenone retention indices and UV-VIS spectra (diode array detection). – J. Chromatography **404**: 195–214.

Frisvad, J. C. & U. Thrane (1993). Liquid chromatography of mycotoxins. – In: Betina, V. (Eds). Chromatography of mycotoxins: techniques and applications. J. Chromatography Library **54**: 253–372.

Fröhlich, J., K. D. Hyde & O. Petrini (2000). Endophytic fungi associated with palms. – Mycological Research **104**: 1202–1212.

Fungi Canadenses No. 38: *Monilinia fructicola*.

Gams, W. (1971a). *Cephalosporium*-artige Hyphomyceten. – Fischer, Stuttgart, 262 pp.

Gams, W. (1971b). *Tolypocladium*, eine Hyphomycetengattung mit geschwollenen Phialiden. – Persoonia **6**: 185–191.

Gams, W. (1976). Some new or noteworthy species of *Mortierella*. – Persoonia **9**: 111–144.

Gams, W. (1977). A key to the species of *Mortierella*. – Persoonia **9**: 381–391.

Gams, W. (Hrsg., 2006). *Hypocrea* and *Trichoderma* studies marking the 90th birthday of Joan M. Dingley. – Studies in Mycology **56**: 1–179.

Gams, W. & R. Zare (2002). New generic concepts in *Verticillium* sect. *Prostrata*. – Mycological Research **106**: 130–131.

Gerlach, W. (1959a). Beiträge zur Kenntnis der Gattung *Cylindrocarpon* Wr. II. *C. victoriae*. – Phytopath. Z. **35**: 292–300.

Gerlach, W. (1959b). Beiträge zur Kenntnis der Gattung *Cylindrocarpon* Wr. III. *C. olidum* Wr. und seine phytopathologische Bedeutung. – Phytopath. Z. **35**: 333–346.

Gerlach, W. & H. Nirenberg (1982). The genus *Fusarium* – a pictorial atlas. – Mitteilungen aus der biologischen Bundesanstalt für Land- und Forstwirtschaft Berlin-Dahlem **209**: 1–406.

Gilgado, F., C. Serena, J. Cano, J. Gene & and J. Guarro (2006). Antifungal susceptibilities of the species of the *Pseudallescheria boydii* complex. – Antimicrob. Agents Chemother. **50**: 4211-4213.

Gräser, Y., J. Scott, & R. Summerbell (2008). The new species concept in dermatophytes – a polyphasic approach. – Mycopathologia **166**: 239–256.

Graham, J. H. & E. S. Luttrell (1961). Species of *Leptosphaerulina* on forage plants. – Phytopathology **51**: 650–693.

Gravesen, S., J. C. Frisvad & R. A. Samson (1994). Microfungi. – Munksgaard, Copenhagen, 168 pp.

Gravesen, S., P. A. Nielsen, R. Iversen & K. F. Nielsen (1999). Microfungal contamination of damp buildings – examples of risk constructions and risk materials. – Environmental Health Perspectives **107**: 505–508.

Greenacre, M. J. (1986). SimCA: a program to perform simple correspondence analysis. – Amer. Statistician **40**: 230–231.

Greenacre, M. J. (1989). Theory and applications of correspondence analysis. – Academic Press, London, U. K., 3. Ed., 364 pp.

Greenacre, M. J. (1993). Correspondence analysis in practice. – Academic Press, London, U.K.

Gregory, P. H. (1966). The fungus spore: what it is and what is does. – In: Madelin, M. F. (Eds). The fungus spore. Butterworth, London: 1–13.

Guarro, J., A. S. Kantarcioglu, R. Horre, J. L. Rodriguez-Tudela, M. Cuenca Estrella, J. Berenguer & G. S. de Hoog (2006). *Scedosporium apiospermum*: changing clinical spectrum of a therapy-refractory opportunist. – Med. Mycol. **44**: 295-327.

Hämmerli, U. A., U. E. Brändle, O. Petrini & J. M. McDermott (1992). Differentiation of *Discula umbrinella* (teleomorph: *Apiognomonia errabunda*) from beech, chestnut, and oak using randomly amplified polymorphic DNA markers. – Mol. Plant-Microbe Interact. **5**: 479–483.

Hawksworth, D. L. (1971). A revision of the genus *Ascotricha*. – Mycol. Papers **126**: 1–28.

Hawksworth, D. L. (2006). The fungal dimension of biodiversity: magnitude, significance, and conservation. – Mycological Research **95** (6): 641–655. doi:10.1016/S0953-7562(09)80810-1.

Hawksworth, D. L. & J. I. Pitt (1983). A new taxonomy for *Monascus* species based on cultural and microscopical characters. – Austr. J. Bot. **31**: 51–61.

Hawksworth, D. L. & H. Y. Yip (1981). *Coniochaeta angustispora* sp. nov. from roots in Australia, with a key to the species known in culture. – Austr. J. Bot. **29**: 377–384.

Hermanides-Nijhof, E. J. (1977). *Aureobasidium* and allied genera. – Studies in Mycology **15**: 141–177.

Hermet, A., D. Méheust, J. Mounier, G. Barbier & J.-L. Jany (2012). Molecular systematics in the genus *Mucor* with special regards to species encountered in cheese. – Fungal Biology **116**: 692–705.

Hesseltine, C. W. & P. Anderson (1956). The genus *Thamnidium* and a study of the formation of its zygospores. – Am. J. Bot. **43**: 696–702.

Hesseltine, C. W., C. R. Benjamin & B. S. Mehrotra (1959). The genus *Zygorhynchus*. – Mycologia **51**: 173–194.

Hibbett, D. S., M. Binder, J. F. Bischoff, M. Blackwell, P. F. Cannon, O. E. Eriksson et al. (2007). A higher-level phylogenetic classification of the Fungi. – Mycological Research **111**: 509–47.

Hirooka, Y., A. Y. Rossman & P. Chaverri (2011). A morphological and phylogenetic revision of the *Nectria cinnabarina* species complex. – Studies in Mycology **68**: 35–56.

Ho, M. H.-M., F. M. Dugan & S. C. Jong (1999). *Cladosporium* and *Cladophora* in culture: descriptions and an expanded key. – Mycotaxon **72**: 115–157.

Hsieh, W. H. (1979). The causal organism of sugarcane leaf blight. – Mycologia **71**: 892–898.

Jaklitsch, W. (2009). European species of *Hypocrea*. Part I. The green-spored species. – Studies in Mycology **63**: 1–91.

Jaklitsch, W. M., C. P. Kubicek & I. S. Druzhinina (2008). Three European species of *Hypocrea* with reddish brown stromata and green ascospores. – Mycologia **100**: 796–815.

Jaklitsch, W. M., G. J. Samuels, S. J. Dodd, B.-S. Lu & I. S. Druzhinina (2006). *Hypocrea rufa / Trichoderma viride*: a reassessment, and description of five closely related species with and without warted conidia. – Studies in Mycology **56**: 135–177.

Jarvis, W. R. (1977). *Botryotinia* and *Botrytis* species: taxonomy, physiology and pathogenicity. – Res. St. Can. Dep. Agric. Harrow, Monogr. **15**: 1–195.

Joffe, A. Z. (1986). *Fusarium* species: their biology and toxicology. – John Wiley & Sons, Inc. New York, Chichester, Brisbane, Toronto, Singapore, 588 pp.

Johnston, A. & C. Booth (1983). Plant pathologist's pocketbook. – 2nd Ed., Commonwealth Mycological Institute, Kew. Surrey, 439 pp.

Jong, S. C. & G. F. Yuan (1985). *Actinomucor taiwanensis* sp. nov., for manufacture of fermented soybean food. – Mycotaxon **23**: 261–264.

Kendrick, B. (1992). The fifth Kingdom. – 2nd Ed. Ontario, Mycol. Publications, 406 pp.

Kiffer, E. & M. Morelet (1997). Les deuteromycètes: classification et clés d'identification générique. – INRA, Paris, 306 pp.

Kiffer, E. & M. Morelet (1999). The Deuteromycetes, mitosporic fungi: classification and generic keys. – Science Publishers, Inc., Enfield, NH, USA, 273 pp.

Kindermann, J., Y. El-Ayouti, G. J. Samuels & C. P. Kubicek (1998). Phylogeny of the genus *Trichoderma* based on sequence analysis of the internal transcribed spacer region 1 of the rDNA cluster. – Fungal Genet. Biol. **24**: 298–309.

Kirk, P. M., P. F. Cannon, J. A. Stalpers & D. W. Minter (2011). Dictionary of the fungi. – 10th Ed. Commonwealth Scientific and Industrial Research Organisation (CSIRO), 784 pp.

Klein, D. & D. E. Eveleigh (1998). Ecology of *Trichoderma*. – In: Harman, G. E. & C. P. Kubicek (eds). *Trichoderma* and *Gliocladium*. Taylor & Francis, London: 57–74.

Klich, M. A. (1993). Morphological studies of *Aspergillus* section *Versicolores* and related species. – Mycologia **85**: 100–107.

Klich, M. A. (2002). Identification of common *Aspergillus* species. – Centraalbureau voor Schimmelcultures, Utrecht, 116 pp.

Kohn, L. M. (1992). Developing new characters for fungal systematics: an experimental approach for determining the rank of resolution. – Mycologia **84**: 139–153.

Komon-Zelazowska, M., J. Bissett, D. Zafari, L. Hatvani, L. Manczinger, S. Woo, M. Lorito, L. Kredics, C. P. Kubicek & I. S. Druzhinina (2007). Genetically closely related but phenotypically divergent *Trichoderma* species cause green mold disease in oyster mushroom farms worldwide. – Appl. Environ.Microbiol. **73**: 7415–7426.

Kozakiewicz, Z. (1989). *Aspergillus* species on stored products. – Mycological Papers **161**: 1–188.

Kraus, G. F., I. S. Druzhinina, W. Gams, J. Bissett, D. Zafari, G. Szakacs, A. Koptchinski, H. Prillinger, R. Zare & C. P. Kubicek (2004). *Trichoderma brevicompactum* sp. nov. – Mycologia **96**: 1059-1073.

Kubiček, C. P. & G. E. Harman (Eds) (1998). *Trichoderma* and *Gliocladium*. – Tayler & Francis, London, Vol. I: 278 pp.; Vol. II: 393 pp.

Kubiček, C. P., M. Komon Zelazowska & I. S. Druzhinina (2008). Fungal genus *Hypocrea / Trichoderma*: from barcodes to biodiversity. – J. Zhejiang Univ. Sci. **B9**: 753–763.

Kuhls, K., E. Lieckfeldt, G. J. Samuels, W. Meyer, C. P. Kubicek & T. Börner(1997). Revision of *Trichoderma* sect. *Longibrachiatum* including related teleomorphs based on analysis of ribosomal DNA internal transcribed spacer sequences. – Mycologia **89**: 442–460.

Leslie, J. F., B. A. Summerell (2006). The *Fusarium* laboratory manual. – Blackwell Publishing, Oxford, U. K., 400 pp.

Leuchtmann, A., O. Petrini, L. E. Petrini & G. C. Carroll (1992). Isozyme polymorphism in six endophytic *Phyllosticta* species. – Mycological Research **96**: 287–294.

Liu, X.-Y., Huang H. & Zheng R.-Y. (2007). Molecular phylogenetic relationships within *Rhizopus* based on combined analyses of ITS rDNA and pyrG gene sequences. – Sydowia **59** (2): 235–253.

Lombard, L., P. W. Crous, B. D. Wingfield & M. J. Wingfield (2010). Systematics of *Calonectria*: a genus of root, shoot and foliar pathogens. – Studies in Mycology **66**: 1–71.

Lu, B., I. S. Druzhinina, P. Fallah, P. Chaverri, C. Gradinger, C. P. Kubicek & G. J. Samuels (2004). *Hypocrea/Trichoderma* species with *Pachybasium*-like conidiophores: teleomorphs for *T. minutisporum* and *T. polysporum* and their newly discovered relatives. – Mycologia **96**: 310–342.

Ludwig, J. A. & J. F. Reynolds (1988). Statistical ecology. A primer on methods and computing. – J. Wiley and Sons, New York, USA, 337 pp.

Lundqvist, N. (1972). Nordic *Sordariaceae* s. lat. – Symbolae Botanicae Upsalienses **20**: 1–374.

Lutzoni, F. & R. Vilgalys (1995). Integration of morphological and molecular data sets in estimating fungal phylogenies. – Can. J. Bot. **73** (1): S649–S659.

Malloch, D. & R. F. Cain (1972). The *Trichocomataceae*: Ascomycetes with *Aspergillus*, *Paecilomyces*, and *Penicillium* states. – Can. J. Bot. **50**: 2613–2628.

Malloch, D. & R. F. Cain (1973). The genus *Thielavia*. – Mycologia **65**: 1055–1077.

Marasas, W. F. O., P. E. Nelson & T. A. Toussoun (1984). Toxigenic *Fusarium* species. Identity and mycotoxicology. – The Pennsylvania State University Park. University Park and London, 328 pp.

Marklein, G., M. Josten, U. Klanke, E. Muller, R. Horre, T. Maier, et al. Matrix-assisted laser desorption ionization-time of flight mass spectrometry for fast and reliable identification of clinical yeast isolates. – J. Clin. Microbiol. **47** (9): 2912–2917.

Martin, G. S., D. M. Mannino, S. Eaton & M. Moss (2003). The epidemiology of sepsis in the United States from 1979 through 2000. – N. Engl. J. Med. **348**: 1546–1554.

McLaughlin, D. J., M. E. Berres & L. J. Szabo (1995). Molecules and morphology in basidiomycete phylogeny. – Can. J. Bot. **73**: S684–S692.

Mel'nik, V. A. (1977). Key to fungi of the genus *Ascochyta* Lib. – Leningrad, Russland, 245 pp. (In Russian).

Monod, M. (1983). Monographie taxonomique des *Gnomoniaceae*. – Beihefte zur Sydowia **9**: 1–315.

Monte, E., P. D. Bridge & B. C. Sutton (1990). Physiological and biochemical studies in Coelomycetes. *Phoma*. – Studies in Mycology **32**: 1–28.

Moreau-Froment, M. (1956). Les *Neurospora*. – Bull. Soc. Bot. Fr. **103**: 678–738.

Morton, F. J. & G. Smith (1963). The genera *Scopulariopsis* Bainier, *Microascus* Zukal, and *Doratomyces* Corda. – Mycological Papers **86**: 1–96.

Mücke, W. & Ch. Lemmen (2004). Schimmelpilze. Vorkommen – Gesundheitsgefahren – Schutzmassnahmen. – Ecomed, 3rd edition, 182 pp.

Müller, E. & W. Löffler (1992). Mykologie. – Thieme, Stuttgart, 5th ed. 367 pp.

Nag Raj, T. R. (1993).Coelomycetous anamorphs with appendage-bearing conidia. – Mycologue Publications, Waterloo, 1101 pp.

Nelson, P. E., T. A. Toussoun & W. F. O. Marasas (1983). *Fusarium* species. An illustrated manual for identification. – The Pennsylvania State University Park. University Park and London, 193 pp.

Nielsen, K. F., S. Gravesen, P. A. Nielsen et al. (1999). Production of Mycotoxins on artificially and naturally infested building materials. – Mycopathologia **145**: 43–56.

Nielsen, K. F., P. A. Nielsen & G. Holm (2000). Growth of moulds on building materials under different humidities. – Proceedings of Healty Buildings 2000, Espoo **3**: 283–288.

Nielsen, K. F., M. Østergaard, T. O. Larsen & U. Thrane (1998a). Production of mycotoxins on water damaged gypsum boards. – International Biodeterioration & Biodegradation **42**: 1–7.

Nielsen, K. F., U. Thrane, T. O Larsen et al. (1998b). Production of mycotoxins on artificially inoculated building materials. – International Biodeterioration & Biodegradation **42**: 8–15.

Okada, G., K. Jacobs, T. Kirisits et al. (2000). Epitypification of *Graphium penicilloides* Corda with comments on the phylogeny and taxonomy of *Graphium*-like fungi. – Studies in Mycology **45**: 169–188.

Okada, G., K. A. Seifert, A. Takematsu et al. (1998). A molecular phylogenetic reappraisal of the *Graphium* complex based on 18S rDNA sequences. – Can. J. Bot. **76**: 1495–1506.

Orr, G. F. (1979). The genus *Pseudogymnoascus*. – Mycotaxon **8**: 165–173.

Orr, G. F., H. H. Kuehn & O. A. Plunkett (1963). The genus *Gymnoascus* Baranetzky. – Mycopathol. Mycol. Appl. **21**: 1–18.

Palm, M. E., W. Gams & H. I. Nirenberg (1995). *Plectosporium*, a new genus for *Fusarium tabacinum*, the anamorph of *Plectosphaerella cucumerina*. – Mycologia **87**: 397–406.

Pankhurst, R. J. (1991). Practical taxonomic computing. – Cambridge University Press, Cambridge, U.K., 202 pp.

Peberdy, J. F. (Hrsg., 1987). *Penicillium* and *Acremonium*. – Plenum Press, New York und London, 297 pp.

Peerally, A. (1991). The classification and phytopathology of *Cylindrocladium* species. – Mycotaxon **40**: 323–366.

Perkins, D. D., A. Radford & M. S. Sachs (2000). The *Neurospora* Compendium. – Academic Press, London, UK, 350 pp.

Petrini, L. E. (1992). *Rosellinia* species of the temperate zones. – Sydowia **44**: 169–281.

Petrini, L. E., O. Petrini, A. Leuchtmann & G. C. Carroll (1991). Conifer inhabiting species of *Phyllosticta*. – Sydowia **43**: 148–169.

Petrini, L. E., O. Petrini & T. N. Sieber (1987). Host specificity of *Hypoxylon fuscum*: A statistical approach to the problem. – Sydowia **40**: 227–234.

Petrini, O. (1986). Taxonomy of endophytic fungi of aerial plant tissues. – In: Fokkema, N. J. & J. van den Heuvel (Eds). Microbiology of the phyllosphere. Cambridge University Press: 175–187.

Petrini, O. & L. E. Petrini (1996). Polyphasische Taxonomie: Probleme und Methoden. – Mycologia Helvetica **8** (2): 83–90.

Petrini, O., L. E. Petrini, G. Laflamme & G. B. Ouellette (1989). Taxonomic position of *Gremmeniella abietina* and related species: a reappraisal. – Can. J. Bot. **76**: 2805–2814.

Petrini, O. & T. N. Sieber (2000). Computer assisted taxonomy and documentation. Chapter 8. – In: McLaughlin, D. J., E. G. Mclaughlin & P. A. Lemke (Eds). The Mycota VII. Springer-Verlag Berlin Heidelberg: 203–215.

Petrini, O., J. Stone & F. E. Carroll (1982). Endophytic fungi in evergreen shrubs in Western Oregon: a preliminary study. – Can. J. Bot. **60**: 789–796.

Pfaller, M. A. & D. J. Diekema (2010). Epidemiology of invasive mycoses in North America. – Crit. Rev. Microbiol. **36**:1–53.

Pitt, J. I. (1974). A synoptic key to the genus *Eupenicillium* and to sclerotigenic *Penicillium* species. – Can. J. Bot. **52**: 2231–2236.

Pitt, J. I. (1979). The genus *Penicillium* and its teleomorphic states *Eupenicillium* and *Talaromyces*. – Academic Press, London, 634 pp.

Pitt, J. I. (2000). A laboratory guide to common *Penicillium* species. – CSIRO Division of food processing, Sydney, 3rd ed., 197 pp.

Pitt, J. I. & A. D. Hocking (1983). *Xeromyces*. – CSIRO Food Res. Q. **42**: 1.

Pitt, J. I. & R. A. Samson (1993). Species names in current use in the *Trichocomaceae* (Fungi: Eurotiales). – Regnum Vegetabile **128**: 13–57.

Powell, K.A., A. Renwick & J. F. Peberdy (Eds, 1994). The genus *Aspergillus*: from taxonomy and genetics to industrial application. – Plenum Press, New York, London, 394 pp.

Prusky, D., S. Freeman & M. Dickman (2000). *Colletotrichum*: host specificity, pathology, and host-pathogen interaction. – APS Press, St. Paul, Minnesota, USA, 400 pp.

Pugh, G. J. F., J. P. Blakeman & G. Morgan-Jones (1964). *Thermomyces verrucosus* and *T. lanuginosus*. – Trans. Br. Mycol. Soc. **47**: 115–121.

Punithalingam, E. (1974). Studies on Sphaeropsidales in culture II. – Mycological Papers **136**: 1–63.

Punithalingam, E. (1979). Graminicolous *Ascochyta* species. – Mycological Papers **142**: 1–214.

Punithalingam, E. (1981). Studies on Sphaeropsidales in culture III. – Mycological Papers **149**: 1–42.

Punithalingam, E. (1988). *Ascochyta* II. Species on monocotyledons (excluding grasses), cryptogams and gymnosperms. – Mycological Papers **159**: 1–235.

Ramirez, C. (1982). Manual and atlas of the *Penicillia*. – Elsevier Biomedical, New York and Oxford, 874 pp.

Rao, V. & G. S. De Hoog (1975). Some notes on *Torula*. – Persoonia **8**: 199–206.

Rao, V. & G. S. De Hoog (1986). New or critical hyphomycetes from India. – Studies in Mycology **28**: 1–84.

Raper, K. B. & D. I. Fennell (1965). The genus *Aspergillus*. – Williams & Wilkins, Baltimore, 686 pp.

Rifai, M. A. (1969). A revision of the genus *Trichoderma*. – Mycological Papers **116**: 1–56.

Robert,V., G. Stegehuis & J. Stalpers (2005). The MycoBank engine and related databases. – http://www.mycobank.org

Rogers, S. O., J. M. McKemy & C. J. K. Wang (1999). Molecular assessment of *Exophiala* and related hyphomycetes. – Studies in Mycology **43**: 122–133.

Rossman, A. Y. (1983). The phragmosporous species of *Nectria* and related genera. – Mycological Papers **150**: 1–164.

Rossman, A. Y. (1989). A synopsis of the *Nectria cinnabarina*-group. – Memoirs of the New York Botanical Garden **49**: 253–265.

Rossman, A. Y., G. J. Samuels, C. T. Rogerson & R. Lowen (1999). Genera of *Bionectriaceae*, *Hypocreaceae* and *Nectriaceae* (Hypocreales, Ascomycetes). – Studies in Mycology **42**: 1–248.

Rotem, J. (1994). The genus *Alternaria*, biology, epidemiology and pathogenicity. – APS Press, St. Paul, Minnesota, USA, 326 pp.

Samson, R. A. (1969). Revision of the genus *Cunninghamella* (Fungi, Mucorales). – Proc. K. ned. Akad. Wet., C. **72**: 322–335.

Samson, R. A. (1972). Notes on *Pseudogymnoascus*, *Gymnoascus* and related genera. – Acta Bot. Neerl. **21**: 517–527.

Samson, R. A. (1974). *Paecilomyces* and some allied hyphomycetes. – Studies in Mycology **6**: 1–119.

Samson, R. A. (1979). A compilation of the Aspergilli described since 1965. – Studies in Mycology **18**: 1–38.

Samson, R. A. (1992). Current taxonomic schemes of the genus *Aspergillus* and its teleomorphs. – In: Bennett, J. W. & M. A. Klich (eds.). *Aspergillus*: biology and industrial applications. Butterworth-Heinemann, Stoneham, Massachusetts: 335–390.

Samson, R. A. (1994). Current systematics of the genus *Aspergillus*. – In: Powell, K. A., A. Renwick & J. F. Peberdy (eds.). The genus *Aspergillus*: from taxonomy and genetics to industrial application. Plenum Press, London: 261–276.

Samson, R. A. & J. C. Frisvad (2004). *Penicillium* subgenus *Penicillium*: new taxonomic schemes, mycotoxins and other extrolites. – Studies in Mycology **49**: 1–260.

Samson, R. A. & W. Gams (1985). Typification of the species of *Aspergillus* and associated teleomorphs. – In: Samson, R. A. & J. I. Pitt (eds.). Advances in *Penicillium* and *Aspergillus* systematics. Plenum Press, New York, London: 31–54.

Samson, R. A. & J. I. Pitt (eds., 1985). Advances in *Penicillium* and *Aspergillus* systematics. – Plenum Press, New York, London, 483 pp.

Samson, R. A. & J. I. Pitt (eds., 1990). Modern concepts in *Penicillium* and *Aspergillus* classification. – Plenum Press, New York, London, 478 pp.

Samson, R. A. & J. I. Pitt (eds., 2000). Integration of modern taxonomic methods for *Penicillium* and *Aspergillus* classification. – Harwood Academic Publishers, Amsterdam, 510 pp.

Samson, R. A. & J. Varga (eds., 2007). *Aspergillus* systematics in the genomic aera. – Studies in Mycology **59**: 1–206.

Samson, R. A., B. Flannigan, M. E. Flannigan et al. (1994). Health implications of fungi in indoor environments. – Elsevier, Amsterdam, 602 pp.

Samson, R. A., A. D. Hocking, J. I. Pitt & A. D. King (1992). Modern methods in food mycology. – Developments in Food Science **31**: 1–388.

Samson, R. A., E. S. Hoekstra & J. C. Frisvad (eds., 2004). Introduction to food- and airborne fungi. 7th edition. – Centraalbureau voor Schimmelcultures, Utrecht, 389 pp.

Samson, R. A., S. Hong, S. W. Peterson, J. C. Frisvad & J. Varga (2007). Polyphasic taxonomy of *Aspergillus* section *Fumigati* and its teleomorph *Neosartorya*. – Studies in Mycology **59**: 147–203.

Samson, R. A. & J. Houbraken (Eds.) (2011). Phylogenetic and taxonomic studies on the genera *Penicillium* and *Talaromyces*. – Studies in Mycology **70** (1): 1–183.

Samson, R. A., J. Houbraken, U. Thrane, J. C. Frisvad & B. Anderson (2010). CBS Laboratory Manual Series 2: Food and indoor fungi. – Centraalbureau voor Schimmelcultures, Utrecht, 390 pp.

Samson, R. A., P. V. Nielsen & J. C. Frisvad (1990). The genus *Neosartorya*: differentiation by scanning electron microscopy and mycotoxin profiles. – In: Samson, R. A. & J. I. Pitt (eds.). Modern concepts in *Penicillium* and *Aspergllus* clasification. Plenum Press, New York, London: 455–467.

Samson, R. A., A. C. Stolk & R. Hadlok (1976). Revision of the subsection fasciculata of *Penicillium* and some allied species. – Studies in Mycology **11**: 1–47.

Samson, R. A., J. Varga & J. Frisvad (Eds.) (2011). Taxonomic studies on the genus *Aspergillus*. – Studies in Mycology **69**: 1–97.

Samson, R. A., N. Yilmaz, J. Houbraken, H. Spierenburg, K. A. Seifert, S. W. Peterson, J. Varga & J. C. Frisvad (2011). Phylogeny and nomenclature of the genus *Talaromyces* and taxa accommodated in *Penicillium* subgenus *Biverticillium*. – Studies in Mycology **70**: 159–183.

Samuels, G. J. (1976). A revision of the fungi formerly classified as *Nectria* subgenus *Hyphonectria*. – Memoirs of the New York Botanical Garden **26**: 1–126.

Samuels, G. J. (1988). Species of *Nectria* (Ascomycetes, Hypocreales) having orange perithecia and colourless, striate ascospores. – Brittonia **40**: 306–331.

Samuels, G. J. (1989). *Nectria* and *Sesquicillium*. – Memoirs of the New York Botanical Garden **49**: 266–285.

Samuels, G. J. (2006). *Trichoderma*: systematics, the sexual state, and ecology. – Phytopathology **69**: 195–206.

Samuels, G. J. & A. Ismaiel (2009). *Trichoderma evansii* and *T. lieckfeldtiae*: two new *T. hamatum*-like species. – Mycologia **101**: 142–156.

Samuels, G. J. & K. A. Seifert (1995). The impact of molecular characters on systematics of filamentous ascomycetes. – Annu. Rev. Phytopathol. **33**: 37–67.

Samuels, G. J., S. L. Dodd, B.-S. Lu, O. Petrini, H.-J. Schroers & I. S. Druzhinina (2006). The *Trichoderma koningii* aggregate species. – Studies in Mycology **56**: 67–133.

Samuels, G. J., S. L. Dodd, W. Gams, L. A. Castelbury & O. Petrini (2002). *Trichoderma* species associated with the green mold epidemic of commercially grown *Agaricus bisporus*. – Mycologia **94**: 146–170.

Samuels, G.J., A. Ismaiel, M. C. Bon, S. De Respinis & O. Petrini (2010). *Trichoderma asperellum sensu lato* consists of two cryptic species. – Mycologia **102** (4): 944–966. DOI:10.3852/09-243.

Samuels, G. J., E. Lieckfeldt & H. I. Nirenberg (1999). *Trichoderma asperellum*, a new species with warted conidia, and redescription of *T. viride*. – Sydowia **51**: 71–88.

Samuels, G. J., O. Petrini, K. Kuhls, E. Lieckfeldt & C. P. Kubicek (1998). The *Hypocrea schweinitzii* complex and *Trichoderma* sect. *Longibrachiatum*. – Studies in Mycology **41**: 1–54.

Samuels, G. J., A. Y. Rossman, R. Lowen & C. T. Rogerson (1991). A synopsis of *Nectria* subgen. *Dialonectria*. – Mycological Papers **164**: 1–48.

Schipper, M. A. A. (1978a). 1. On certain species of *Mucor* with a key to all accepted species. – Studies in Mycology **17**: 1–52.

Schipper, M. A. A. (1978b). 2. On the genera *Rhizomucor* and *Parasitella*. – Studies in Mycology **17**: 53–71.

Schipper, M. A. A. (1984). The *Rhizopus stolonifer*-group and *Rh. oryzae*. – Studies in Mycology **25**: 1–19.

Schipper, M. A. A. & J. A. Stalpers (1984). The *Rhizopus microsporus*-group. – Studies in Mycology **25**: 20–34.

Schol–Schwarz, M. B. (1970). Revision of the genus *Phialophora* (Moniliales). – Persoonia **6**: 59–94.

Schroers, H.-J. (2001). A monograph of *Bionectria* (Ascomycota, Hypocreales, *Bionectriaceae*) and its *Clonostachys* anamorphs. – Studies in Mycology **46**: 1–214.

Schroers, H.-J., G. J. Samuels, K. A. Seifert & W. Gams (1999). Classification of the mycoparasite *Gliocladium roseum* in *Clonostachys* as *C. rosea*, its relationship to *Bionectria ochroleuca*, and notes on other *Gliocladium*-like fungi. – Mycologia **91**: 365–385.

Schubert K., J. Z. Groenewald, U. Braun, J. Dijksterhuis, M. Starink, C. F. Hill, P. Zalar, G. S. de Hoog & P. W. Crous (2007). Biodiversity in the *Cladosporium herbarum* complex (*Davidiellaceae*, Capnodiales), with standardisation of methods for *Cladosporium* taxonomy and diagnostics. – Studies in Mycology **58**: 105–156.

Seifert, K. A. (1985). A monograph of *Stilbella* and some allied Hyphomycetes. – Studies in Mycology **27**: 1–235.

Seifert, K., G. Morgan-Jones, W. Gams, & B. Kendrick (2011). The Genera of Hyphomycetes. – CBS Biodiversity Series **9**: 1–997.

Seifert, K. A. & G. Okada (1993). *Graphium* anamorphs of *Ophiostoma* and similar anamorphs of other ascomycetes. – In: Wingfield, M. J., K. A. Seifert & J. F. Webber (eds.). *Ceratocystis* and *Ophiostoma*. APS Press, 25–42.

Seifert, K. A., R. A. Samson, J. R. Dewaard, J. Houbraken, C. A. Levesque, J. M. Moncalvo, G. Louis-Seize & P. D. N. Hebert (2007). Prospects for fungus identification using CO1 DNA barcodes, with *Penicillium* as a test case. – Proceedings of the National Academy of Sciences U.S.A. **104**: 3901–3906.

Seifert, K. A., B. D. Wingfield & M. J. Wingfield (1995). A critique of DNA sequence analysis in the taxonomy of filamentous Ascomycetes and ascomycetous anamorphs. – Can. J. Bot. **73** (Suppl. 1): S760–S767.

Seifert, K. A., F. Sieber-Canavesi, O. Petrini, A. K. M. Ekramoddoullah & C. E. Dorworth (1991). Characterization of Canadian and European *Melanconium* from some *Alnus* species by morphological, cultural, and biochemical studies. – Can. J. Bot. **69**: 2170–2176.

Seng, P., M. Drancourt, F. Gouriet, B. La Scola, P. E. Fournie, J. M. Rolain et al. (2009). Ongoing revolution in bacteriology: routine identification of bacteria by matrix-assisted laser desorption ionization time-of-flight mass spectrometry. – Clinical Infectious Diseases **49** (4): 543–551.

Sieber, T. N., O. Petrini & M. J. Greenacre (1998). Correspondence analysis as a tool in fungal taxonomy. – Syst. Appl. Microbiol. **21**: 433–441.

Sieber-Canavesi, F., O. Petrini & T. N. Sieber (1991). Endophytic *Leptostroma* species on *Picea abies*, *Abies alba*, and *Abies balsamea*: a cultural, biochemical, and numerical study. – Mycologia **83**: 89–96.

Sigler, L. & J. W. Carmichael (1976). Taxonomy of *Malbranchea* and some other hyphomycetes with arthroconidia. – Mycotaxon **4**: 349–488.

Simmons, E. G. (1967). Typification of *Alternaria*, *Stemphylium* and *Ulocladium*. – Mycologia **59**: 67–92.

Simmons, E. G. (1969). Perfect states of *Stemphylium*. – Mycologia **61**: 1–26.

Simmons, E. G. (1985). Perfect states of *Stemphylium* II. – Sydowia **38**: 284–293.

Simmons, E. G. (1996a). *Alternaria* themes and variations (145–149). – Mycotaxon **57**: 391–409.

Simmons, E. G. (1996b). *Alternaria* themes and variations (150). – Mycotaxon **59**: 319–335.

Simmons, E. G. (1997). *Alternaria* themes and variations (151–223). – Mycotaxon **65**: 1–91.

Simmons, E. G. (1998). *Alternaria* themes and variations (224–225). – Mycotaxon **68**: 417–427.

Simmons, E. G. (2007). *Alternaria*. An identification manual. – CBS Biodiversity series **6**: 1–775.

Singh, K., J. C. Frisvad, U. Thrane & S. B. Mathur (1991). An illustrated manual on identification of some seed-borne *Aspergilli*, *Fusaria*, *Penicillia*, and their mycotoxins. – Danish Government, Institute of Seed Pathology, Hellerup, Denmark, 133 pp.

Sinha, K. K. & D. Bhatnagar (1998). Mycotoxins in agriculture and food safety. – Marcel Dekker, Inc. New York, USA, 536 pp.

Sivanesan, A. (1977). The taxonomy and pathology of *Venturia* species. – Bibliotheca Mycologica **59**: 1–139.

Sivanesan, A. (1984). The bitunicate ascomycetes and their anamorphs. – J. Cramer, Vaduz, 701 pp.

Sivanesan, A. (1987). Graminicolous species of *Bipolaris*, *Curvularia*, *Drechslera*, *Exserohilum* and their teleomorphs. – Mycological Papers **158**: 1–261.

Smiley, R. W., P. H. Dernoeden & B. B. Clarke (2000). Compendium of turfgrass diseases. – APS Press, St. Paul, Minnesota, USA, 4th ed., 98 pp.

Smith, J. E. & R. S. Henderson (1991). Mycotoxins and animal foods. – CRC Press, Inc., Boca Raton, FL, 875 p.

Smith, J. E. & M. O. Moss (1985). Mycotoxins: formation, analysis and significance. – John Wiley & Sons, Chichester, 148 pp.

Sneath, P. H. A. (1989). Analysis and interpretation of sequence data for bacterial systematics: the view of a numerical taxonomist. – Syst. Appl. Microbiol. **12**: 15–31.

Sneath, P. H. A. (1995). Thirty years of numerical taxonomy. – Syst. Biol. **44**: 281–298.

Sneath, P. H. A. & R. R. Sokal (1973). Numerical taxonomy; the principles and practice of numerical classification. – Freeman, San Francisco, 573 pp.

Splittstoesser, D. F., J. M. Lammers, D. L. Downing & J. J. Churey (1989). Heat resistance of *Eurotium herbariorum*, a xerophilic mold. – Journal of food Science **54**: 683–685.

St-Germain, G. & R. Summerbell (2011). Identifying fungi. 2nd Ed. – Star Publishing Company, Belmont, USA, 378 pp.

Stadler, M., H. Wollweber, A. Mühlbauer et al. (2001a). Secondary metabolic profiles, genetic fingerprints and taxonomy of *Daldinia* and allies. – Mycotaxon **77**: 379–429.

Stadler, M., H. Wollweber, A. Mühlbauer et al. (2001b). Molecular chemotaxonomy of *Daldinia* and other Xylariaceae. – Mycological Research **105**: 1191–1205.

Stolk, A. C. (1965). Thermophilic species of *Talaromyces* and *Thermoascus*. – Antonie van Leeuwenhoek **21**: 262–276.

Stolk, A. C. & R. A. Samson (1971). Studies on *Talaromyces* and related genera. I. *Hamigera* gen. nov. and *Byssochlamys*. – Persoonia **6**: 341–357.

Stolk, A. C. & R. A. Samson (1972). The genus *Talaromyces*. – Studies in Mycology **2**: 1–65.

Stolk, A. C. & R. A. Samson (1983). The ascomycetous genus *Eupenicillium* and related *Penicillium* anamorphs. – Studies in Mycology **23**: 1–149.

Streiblová, E. (1988). Cytological methods. – In: Campbell, I. & J. H. Duffus (eds.). Yeast – A practical approach. IRL Press, Washington: 9–49.

Subramanian, C. V. (1972). The perfect states of *Aspergillus*. – Curr. Sci. **41**: 755–761.

Summerbell, R. C., C. Gueidan, H.-J. Schroers, G. S. de Hoog, M. Starink, Y. Arocha Rosete, J. Guarro & J. A. Scott (2011). *Acremonium* phylogenetic overview and revision of *Gliomastix*, *Sarocladium*, and *Trichothecium*. – Studies in Mycology **68**: 139–162.

Summerell, B. A., J. F. Leslie, D. Backhouse, W. L. Bryden & L. W. Burgess (eds., 2001). *Fusarium*: Paul E. Nelson Memorial Symposium. – APS Press, 392 pp.

Sung, G. H., N. L. Hywel-Jones, J. M. Sung, J. J. Luangsard, B. Shrestha & J. W. Spatafora (2007). Phylogenetic classification of *Cordiceps* and the clavicipitaceous fungi. – Studies in Mycology **57**: 5–59.

Sutton, B. C. (1980). The Coelomycetes. – Commonwealth Mycological Institute Kew, Surrey, England, 696 pp.

Sutton, B. C. & I. A. S. Gibson (1977). CMI descriptions of pathogenic fungi and bacteria nos 535, 536, 258.

Svendsen, A. & J. C. Frisvad (1994). A chemotaxonomic study of the terverticillate *Penicillia* based on high-performance liquid chromatography of secondary metabolites. – Mycological Research **98**: 1317–1328.

Theler, A. (1995). Morphological data, molecular data, and total evidence in phylogenetic analysis. – Can. J. Bot. **73** (Suppl. 1): S667–S683.

Thompson, G. R., J. Cadena & T. F. Patterson (2009). Overview of antifungal agents. – Clin. Chest Med. **30**: 203–215.

Tjamos, E. C., R. C. Rowe, J. B. Heale & D. R. Fravel (2000). Advances in *Verticillium*: Research and Disease management. – APS Press, St. Paul, Minnesota, 376 p.

Tonolla M., C. Benagli, S. De Respinis, V. Gaia & O. Petrini (2009). Mass spectrometry in the diagnostic laboratory. – Pipette **3**: 20–25.

Truckess, M. W. & A. E. Pohland (Hrsg., 2000). Mycotoxin protocols. – Methods in molecular biology **157**: 1–256.

Tubaki, K. (1955). Studies on Japanese Hyphomcetes. II. Fungicolous group. – Nagaoa **5**: 11–40.

Tukey, J. W. (1977). Exploratory data analysis. – Adddison-Wesley, Reading, MA, USA, 688 pp.

Udagawa, S. & S. Uchiyama (2002). *Neocarpenteles*: a new ascomycete genus to accommodate *Hemicarpenteles acanthosporus*. – Mycoscience **43**: 3–6.

Uecker, F. A. (1988). A world list of *Phomopsis* names with notes on nomenclature, morphology and biology. – Mycologia Mem. **13**: 1–231.

Uecker, F. A. (1993). Development and cytology of *Plectosphaerella cucumerina*. – Mycologia **85**: 470–479.

Ulloa, M. & R. T. Hanlin (2000). Illustrated dictionary of mycology. – APS Press, St. Paul, Minnesota, USA, 448 pp.

Untereiner, W. A., N. A. Straus & D. Malloch (1995). A molecular-morphotaxonomic approach to the systematics of the *Herpotrichiellaceae* and allied black yeasts. – Mycological Research **99**: 897–913.

van Oorschot, C. A. N. (1980). A revision of *Chrysosporium* and allied genera. – Studies in Mycology **20**: 1–89.

Varga J., M. Due, J. C. Frisvad & R. A. Samson (2007). Taxonomic revision of *Aspergillus* section *Clavati* based on molecular, morphological and physiological data. – Studies in Mycology **59**: 89–106.

Verkley, G. J. M. (1999). A monograph of the genus *Pezicula* and its anamorphs. – Studies in Mycology **44**: 1–180.

von Arx, J. A. (1957). Die Arten der Gattung *Colletotrichum* Corda. – Phytopath. Z. **29**: 413–468.

von Arx, J. A. (1970). A revision of the fungi classified as *Gloeosporium*. – Bibliotheca Mycologica **24**: 1–203.

von Arx, J. A. (1975a). Revision of *Microascus* with the description of a new species. Persoonia **8**: 191–197.

von Arx, J. A. (1975b). On *Thielavia* and some similar genera of ascomycetes. – Studies inMycology **8**: 1–29.

von Arx, J. A. (1981a). On *Monilia sitophila* and some family of ascomycetes. – Sydowia **34**: 13–29.

von Arx, J. A. (1981b). The genera of fungi sporulating in pure culture. – 2nd Ed. Cramer, Vaduz, 424 pp.

von Arx, J. A. (1982). A key to the species of *Gelasinospora*. – Persoonia **11**: 443–449.

von Arx, J. A. (1984), Notes on *Monographella* and *Microdochium*. – Trans. British Mycol. Soc. **83**: 373–374.

von Arx, J. A. (1986). The ascomycete genus *Gymnoascus*. – Persoonia **13**: 173–183.

von Arx, J. A., M. Dreyfuss & E. Müller (1984). A revaluation of *Chaetomium* and the Chaetomiaceae. – Persoonia **12**: 169–179.

von Arx, J. A., J. Guarro & M. J. Figueras (1986). The ascomycete genus *Chaetomium*. – Nova Hedwigia, Beiheft **84**: 1–162.

Warren, C. (1977). Extrinsic allergic alveolitis: a disease commoner in non-smokers. – Thorax **32**(5): 567–569.

Watanabe, T. (2002). Pictorial atlas of soil and seed fungi. Morphologies of cultured fungi and key to species. – CRC Press, Boca Raton, FLA. 2nd ed., 486 pp.

Weber, E. (2002). The *Lecythophora-Coniochaeta* complex: I. Morphological studies on *Lecythophora* species isolated from *Picea abies*. – Nova Hedwigia **74**: 159–185.

Weber, H. (Ed., 1993). Allgemeine Mykologie. – Gustav Fischer Verlag, Jena-Stuttgart, 541 S.

Wechtl, E. E. (1990). *Phomopsis* (Coelomycetes) species on Compositae and Umbelliferae: a critical evaluation of characters with keys. – Linzer Biol. Beitr. **22**: 161–173.

Whalley, A. J. S. & G. N. Greenhalgh (1971). Chemical races of *Hypoxylon rubiginosum*. – Trans. Br. mycol. Soc. **57**: 161–162.

Whalley, A. J. S. & R. L. Edwards (1995). Secondary metabolites and systematic arrangement within the *Xylariaceae*. – Can. J. Bot. **73** (Suppl. 1): S802–S810.

Whalley, A. J. S. & M. A. Whalley. (1977). Stromal pigments and taxonomy of *Hypoxylon*. – Mycopathologia **61**: 99–103.

White, J. F. & G. Morgan-Jones (1983). Studies in the genus *Phoma*. II. Concerning *Phoma sorghina*. – Mycotaxon **18**: 5–13.

White, J. F. & G. Morgan-Jones (1986). Studies in the genus *Phoma*. V. Concerning *Phoma pomorum*. – Mycotaxon **25**: 461–466.

White, J. F. & G. Morgan-Jones (1987). Studies in the genus *Phoma*. VII. Concerning *Phoma glomerata*. – Mycotaxon **28**: 437–445.

Williams, J. G. K., A. R. Kubelik, K. J. Livak, J. A. Rafalski & T. V. Tingey (1990). DNA polymorphism amplified by arbitrary primers are useful as genetic markers. – Nucleic Acids Res. **18**: 6531–6535.

Wilkins, C.L. and O. Lay Jr. (2006). Identification of microorganisms by mass spectrometry. – Chemical Analysis: A series of monographs on analytical chemistry and its applications. Vol 169. Wiley-Interscience.

Wiltshire, S. P. (1947). Species of *Alternaria* on *Brassica*. – Mycol. Pap. **20**: 1–15.

Zalar, P., G. S. de Hoog G, H.-J. Schroers, J. M. Frank, N. Gunde-Cimerman (2005). Taxonomy and phylogeny of the xerophilic genus *Wallemia* (*Wallemiomycetes* and *Wallemiales*, cl. et ord. nov.). – Antonie van Leeuwenhoek **87**: 311–328.

Zalar, P., G. S. de Hoog G, H.-J. Schroers, P. W. Crous, J. Z. Groenewald & N. Gunde-Cimerman (2007). Phylogeny and ecology of the ubiquitous saprobe *Cladosporium sphaerospermum*, with descriptions of seven new species from hypersaline environments. – Studies in Mycology **58**: 157–183.

Zalar, P., C. Gostincar, G. S. de Hoog, V. Ursic, M. Sudhadham & N. Gunde-Cimerman (2008). Redefinition of *Aureobasidium pullulans* and its varieties. – Studies in Mycology **61**: 21–38.

Zambettakis, E. C. (1954). Recherche sur la systématique des „Sphaeropsidales – Phaeodidymae". – Bull. Soc. Mycol. France **70**: 219–350.

Zheng, R.-Y. & G.-Q. Chen (2001). A monograph of *Cunninghamella*. – Mycotaxon **80**: 1–75.

Zheng, R.-Y., G.-Q. Chen, Huang H. & Liu X. Y. (2007). A monograph of *Rhizopus*. – Sydowia **59** (2): 273–372.

Zycha, H., R. Siepmann & G. Linnemann (1969). Mucorales, eine Beschreibung aller Gattungen und Arten dieser Pilzgruppe. – J. Cramer, Lehre, 355 pp.

XIV. Index

A

B

F

G

H

I

J

K

L

M

S

T